T0319413

NARWHALS

Narwhals

〜〜〜

ARCTIC WHALES
in a
MELTING WORLD

A SAMUEL AND ALTHEA STROUM BOOK

TODD MCLEISH

UNIVERSITY OF WASHINGTON PRESS

Seattle and London

This book is published with the assistance of a grant from the Samuel and Althea Stroum Endowed Book Fund.

© 2013 by the University of Washington Press
Printed and bound in the United States of America
Design by Thomas Eykemans
Composed in Warnock Pro, typeface designed by Robert Slimbach
Display type set in Verne Jules, designed by Isaac Tobin
17 16 15 14 13   5 4 3 2 1

All rights reserved. No part of this publication may be reproduced or transmitted in any form or by any means, electronic or mechanical, including photocopy, recording, or any information storage or retrieval system, without permission in writing from the publisher.

UNIVERSITY OF WASHINGTON PRESS
PO Box 50096, Seattle, WA 98145, USA
www.washington.edu/uwpress

LIBRARY OF CONGRESS CATALOGING-IN-PUBLICATION DATA
McLeish, Todd.
Narwhals : Arctic whales in a melting world / Todd McLeish.
    pages cm
Includes bibliographical references and index.
ISBN 978-0-295-99264-8 (pbk : alk. paper)
1. Narwhal.
I. Title.
QL737.C433M43 2013   599.5′43—dc23   2012037685

The paper used in this publication is acid-free and meets the minimum requirements of American National Standard for Information Sciences—Permanence of Paper for Printed Library Materials, ANSI Z39.481984.∞

For Renay.
Once again.

CONTENTS

Prologue   ix

‿‿‿

‿‿‿

PROLOGUE

THE INUIT LEGEND OF THE ORIGIN OF THE NARWHAL HAS BEEN told in many versions throughout the eastern Canadian Arctic, with some being quite long and detailed and others simple and unadorned. Although the basic story is similar, each person brings to the tale slight variations reflecting the individuality of the oral tradition. This version was told to narwhal researcher Martin Nweeia by Elisapee Ootuva, an elder from Baffin Island and the author of the first Inuit-to-English dictionary. In Alaska and the western Canadian Arctic, outside the native range of the narwhal, the story is sometimes called "The Blind Boy and the Loon," and the ending refers to walruses or polar bears.

A wicked woman lived with her daughter and her son, who was born blind. As the son got older, his sight improved, even though the mother tried to convince him of his helpless state. One day a polar bear came near the house and the mother told the son to aim a bow and arrow at the bear through the window covered with seal skin and strike him down. The boy pulled back the arrow and the mother took aim for him. The arrow struck the heart of the bear and although the boy could hear the groans of the dying bear, the mother laughed scornfully at him, saying that he had missed the bear. That night the mother and the daughter had fresh polar bear meat while the mother cooked dog meat for the son. Later the boy's sister told her brother that his shot was successful and secretly gave him some of the polar bear meat.

Time passed and an old man came to the house for a visit. Before he left, he told the young girl how she could help her brother regain his sight. In the spring, he told them to watch for a red-throated loon who would swim trustingly toward them. Once the loon was close enough, the blind brother should wrap his arms around the loon's neck and the loon would take him to the bottom of the lake. Once they came up, his sight would return. The loon told the young man not to tell about his regained sight until later in the summer when he would send a pod of belugas to their campsite.

When summer came and the ice began to break, the belugas began to move. On one occasion, a pod was closer to land than usual. The young man grabbed his harpoon and told his sister to accompany him to help him aim. They went to the shoreline and the mother, seeing the son with a harpoon, became concerned and followed them. Once she was close to them, the son gave the end of the line from the harpoon to his mother, asking her to tie it around her waist to hold the harpooned animal. The concerned mother told her daughter to make sure he was after a small animal as she was tied to the harpoon. The son instead aimed for the largest whale and harpooned him. The mother was cast into the sea. As she submerged she spiraled around the line, with her long hair twisting into a long lance. This is how the narwhal came to be.

NARWHALS

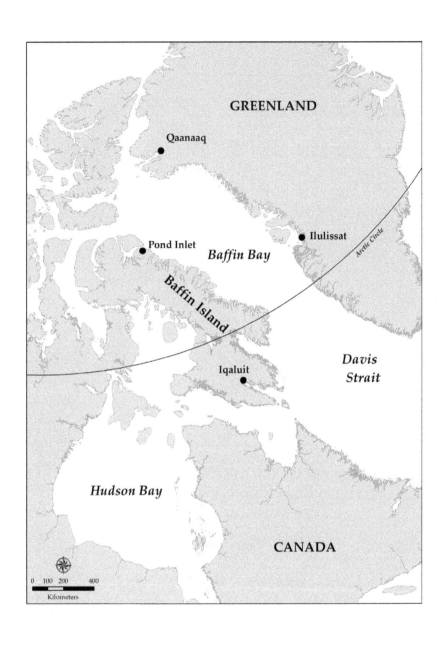

FIRST ENCOUNTER

I THINK I WAS NINE WHEN I FIRST BECAME ENAMORED OF THE narwhal, the mysterious sea creature that most people, even today, aren't sure is actually real. Their confusion arises from the whale's connection to the mythical unicorn, whose horn, modeled after the narwhal's tusk, was reputed to have healing properties. But I knew back then that narwhals were real. And I knew that the truth about the ice whale was even better than the myth with which it had become entangled.

When I stumbled across the narwhal in the pages of the *World Book Encyclopedia*, I was struck by the odd combination of tusk and ice and whale, making the narwhal far more intriguing than the local frogs and turtles this Rhode Island grade-schooler sought out every day. When I discovered a long spiral tusk in a Vancouver curio shop two decades later—labeled with a price tag of $10,000—my interest was rekindled, though I assumed I would never actually see one alive in the wild. Reports of the threats narwhals face from the melting of their icy world, along with conflicting and inaccurate information online and in other venues, prompted me to undertake a series of journeys far above the Arctic Circle to find the truth and reveal the narwhal's unique life cycle and remarkable physical feats.

Narwhals have been surrounded by mystery, mythology, and awe for centuries. They've been celebrated as sea unicorns, held up as proof of the existence of a land-based unicorn, and cherished for their high-quality oils. They've been a vital source of meat in the diet of Arctic natives and important to them as a cultural icon. Today they are surrounded, as well, by great concern for the potential harm that could come to them from retreating sea ice, increased shipping and oil exploration, and changes in environmental conditions caused by a warming planet. For a mammal that is so uniquely adapted to thriving in the numbing waters of the Arctic, warming is a significant threat, especially at the rate that it is taking place today.

Yet it is the narwhal's tusk that is clearly its most distinctive feature: it is the most remarkable tooth in nature, and the trait that has generated the most interest and attention since the earliest days of its discovery. It's the source of its taxonomic name, *Monodon monoceros*, meaning "one tooth, one horn," but its common name comes from its grayish coloration. The Old Norse prefix *Nar* means corpse, and *hval* means whale, so the narwhal is the corpse whale, because its skin color makes it resemble a floating corpse. Despite the origin of its name, Nobel Prize–winning Chilean poet Pablo Neruda wrote that narwhal is "the most beautiful of undersea names, the name of a sea goblet that sings, the name of a crystal spur."

To me, however, it is not just the narwhal's name and tusk that make it distinctive and notable. For narwhals live in a world few can even imagine, where darkness and icy conditions are insurmountable barriers to all but a few predators, some of which may venture north during the warmest months to partake of the region's food-rich waters but then quickly depart for warmer climes at the onset of fall like so many human snowbirds. Their food source is often found more than a mile deep, where few mammals can survive. It's the physiological adaptations to cold and deep diving—and to darkness, too, forcing them to rely much more on their hearing and acoustics than on sight—that put the narwhal on the top of my list of most revered animals.

Equally impressive are their migrations, which carry the largest population of narwhals from the middle of Baffin Bay where they surface in tiny slivers of open water amid the vast expanse of ice—an ecological niche no other mammal can survive in winter—northward to their summering grounds in shallow bays and fjords in the eastern Canadian archipelago and coastal West Greenland. Another population, about which little is known, migrates along the east coast of Greenland, and unknown others range eastward further to the waters north of Norway and Russia. For breeding age females, who give birth in May or June in the early stages of migration after a fourteen-month gestation, the northbound trip is especially daunting. While it is unknown what route they take, most are believed to return year after year to the same locale to raise their young and enjoy the few ice-free months. Scientists have tracked narwhals on their fall trips south, and the animals have been found to follow the same pathways each year. Many of those from Eclipse Sound and Admiralty Inlet near the north end of Baffin Island visit the fjords along Baffin's east coast on their way to their wintering grounds; those summering along the West Greenland coast head straight southwest into deep water; and whales from Somerset Island traverse Lancaster Sound before moving offshore near the northern extent of Baffin Island.

To see large numbers of narwhals as they begin their journey, or during their summer movements back and forth within the fjords, is to witness one of the most impressive wildlife migrations on the planet. Writing about their passage in *Arctic Dreams*, author Barry Lopez was "struck by their agility and swiftness, by the synchronicity of their movements as they swam and dived in unison, and by a quality of alert composure in them, of capability in the face of whatever might happen. Their attractiveness lies partly with their strong, graceful movements in three dimensions, like gliding birds on an airless day."

×   ×   ×

The High Arctic is one of the most difficult places in the world to get to, but it's easy to conjure up what you imagine to be a somewhat accurate picture of what it looks like. Teenage adventure novels, historical accounts of early explorers, and televised nature programs provide plenty of material. In that dreamy world of our imagination, when we're warm and wishing for adventure, the Arctic is teeming with wildlife, the native peoples are determined and self-reliant, the scenery is awe-inspiring, and the ecosystem is clean and healthy. I had spent hours envisioning that world, both as a young man and more recently in middle age, and while those pictures at times mirrored my actual experience of the region, they were often overshadowed by real life, where "harsh" can describe almost everything, from the jagged hillsides to the uncertain flavors. The Arctic is virtually roadless, mostly dark for half the year, ice-covered for longer than that, with cutting winds and blizzard conditions possible at any time. With so few people living on such an enormous land mass, it is easy to feel lonesome there.

It's an environment that's completely alien to me. I'm not a scientist, though sometimes I wish I were. My friends call me a biology groupie because I prefer to spend my free time helping biologists with their research and writing about those experiences. That's part of the reason I chose to undertake the rigors of Arctic travel. But being a groupie in the Arctic can be dangerous, so I tested the waters carefully, taking my first narwhal journey as a tourist and only later joining a team of researchers in Canada and visiting an Inuit hunting camp in Greenland. Along the way I met climate scientists, dental experts, and a ballerina-turned-marine mammalogist, and I helped to moderate an intense disagreement over the purpose of the narwhal's tusk.

The Arctic had captured my imagination years before my first trip there. Having grown up in flat and forested Rhode Island, where the longest vistas on land were across the desolate K-mart parking lot, I fell in love first with the mountains of northern New England, which provided what until then were unimaginably mesmerizing views. Eventually, I dragged my wife, Renay, to explore Alaska, and I haven't been able to keep my mind from wandering back to its powerful beauty again and again. Its mix of craggy mountains, endless rolling tundra, and incomparable and abundant wildlife convinced me that the Arctic was my spiritual home. I wasn't sure I would have the same feelings about the eastern Canadian Arctic, where the narwhal is most eas-ily found, but the sensory and emotional experience of the landscape turned out to be very similar.

The small and scattered population there means that there is little demand for regular commercial travel schedules. But I didn't know that when Renay and I began to make plans for an August trip to the north end of Baffin Island and the tiny Inuit village of Pond Inlet. The first leg of the journey was a flight northeast from Ottawa to Iqaluit, the capital of the newest Canadian province, Nunavut. We rapidly left behind the populated southern rim of Canada and began crossing an immense wilderness, starting with the spectacular boreal forests, where a large portion of North America's songbirds breed among towering spruce, fir, and cedar trees and where eleven percent of the world's carbon is stored. An hour later, when the sky below cleared, we said goodbye to the last tree we would see for ten days and stared down at a brown and rocky terrain that signaled the beginning of a vast expanse of tundra known as the Barren Grounds.

From above, the pockmarked landscape appeared bereft of vege-tation but was certain to be covered in mosses, lichens, grasses, and wildflowers of innumerable varieties that feed the inland subspecies of caribou, Arctic hare, and several types of voles and lemmings. The sun reflected off thousands of puddles and ponds and wetlands of every size that were scattered everywhere, providing breeding grounds for most of the continent's geese, a large percentage of its ducks, and almost all of its shorebirds.

Most notable about the tundra below was what looked like long, straight gouges through the terrain in every direction and extending for as far as I could see, like carved pathways to the horizon. This entire region had once been mountainous, but the expanding and retreating ice sheet had flattened it into a softly rolling plain, leaving behind the abundant wet spots and the harsh scrapes and slashes in the soil.

Straining for a view of the landscape ahead, I eventually saw a major waterway in the distance, Hudson Strait, which drains Hudson Bay and divides northern Quebec from Baffin Island. As we approached, I was dumbfounded to observe what looked like white boats scattered in the water below us. We were too high up and they were too far away, I thought, for them to have been whitecaps, which would have disappeared from view in a few seconds.

The floating objects appeared stationary and reflected the bright sunlight like the hulls of recreational boats. But what were all those boats doing in that virtually uninhabited area? As we got closer, my foolish mistake became clear, another indication that I had entered a realm that was totally unknown to me. The big white boats turned out to be icebergs—some were tall, sharp, and mountainous while others looked more like boxy McMansions—and from our angle it was easy to see their monumental bulk beneath the waterline reflecting a stunning bright blue-green hue. It also became clear that many of the smaller bergs were probably pieces broken off the big ones—bergy bits, as they are sometimes called—as there were distinct trails of smaller chunks of ice leading away from the behemoths. As we edged closer to the Meta Incognita Peninsula in extreme southern Baffin Island, crossed Frobisher Bay, and followed it to Iqaluit, icebergs of all sizes— too numerous to count—built up in bays and inlets and coves, like an explosion of Styrofoam pieces scattered across the landscape. Many were even high and dry on sandy beaches and rocky coastlines, probably pushed on land by winds and the tides and marooned there as the water retreated.

Our flight from Iqaluit to Pond Inlet was aboard a twenty-seat turboprop plane that had extra space between the cabin and cockpit for the storage of cargo. Since almost everything needed for life in the Arctic, from food and clothing to building supplies and vehicles, came from outside the region, most flights carried far more cargo than people.

Before reaching Pond Inlet, we landed in Clyde River, about the mid-way point of Baffin Island and on its far eastern shore. As we approached, the cloud cover opened up to reveal mountains rising up from wide fjords, snow still covering their peaks, and several massive glaciers filling the valleys like smoothly poured cement. At the coast, where the mountains turned to rolling gray-brown hills, north-facing hillsides still showed snow, well hidden from the ever present daylight. Then the clouds returned, hiding the terrain from view once again.

The landing in Clyde River was rougher than I expected—it was my first on a gravel landing strip, and I was happy not to know it was gravel until it was too late to worry. Soon we were in the air again, passing dozens of glacier-covered mountains, fog-filled fjords, a few streams and ponds in the valleys, and sheer cliffs sticking up as sky islands above meringue-froth clouds. Out the opposite window we saw sea ice breaking up like puzzle pieces and maze diagrams, blue-white stained glass windows to the sea below. Our approach to Pond Inlet followed the southern coastline of Eclipse Sound, the wide waterway that separates Baffin Island from Bylot Island, a national park and wildlife refuge to the north. We circled twice around the village to view the three massive icebergs grounded nearby, then landed hard on another gravel runway. After a full day of traveling, we were still a day's boat ride away from our ultimate destination.

Home to 1,300 residents, mostly Inuit, Pond Inlet doesn't look like much—a few dirt roads, mostly government housing, lots of boarded up buildings, and little in the way of public services like hotels, restaurants, or entertainment. Most people travel around the village on four-wheel, all-terrain vehicles, though most homes also have a pickup truck, a snowmobile, and a dogsled parked nearby. Several attractive and colorful public buildings had recently been completed—a library/museum, a health center, and a school—and other building projects were underway, a testament to the more than half-billion dollars in federal support the Canadian government provides annually to the province. But while the village wasn't particularly impressive, the vista from every vantage point certainly was.

The next morning, after struggling to sleep through a night of bright sunshine illuminating our room at a makeshift motel, Renay and I took a stroll around. Early fog quickly burned off to reveal Bylot Island, sixteen miles across the Sound, with a low cloud visible at the water line and stately mountains and glaciers above. As we walked, we saw an impressive variety of blooming flowers—poppies, buttercups, and Arctic cotton among them—scattered among oddly beautiful rocks on hillsides and roadsides. Ravens were everywhere, some seeming to serve as lookouts or guards, others just playing tag, including two perched on an imposing crucifix on the hillside behind the Catholic church. They made a surprising variety of sounds—not just the familiar low croak, but also a crow-like caw, and some odd gurgling noises that sounded more mammal-like than bird. Glaucous gulls and northern fulmars flew up and down the Sound, a few land birds—snow bunting, Lapland longspur, American pipit, and horned lark—turned up in unexpected numbers in town, and an occasional red-throated loon was conspicuous in the water.

Most impressive were the icebergs. Three big ones were grounded in Eclipse Bay, standing guard proudly when illuminated by the bright sunlight, pockmarked and striated from changing water lines. We walked down the beach to the west for a mile, hoping to get a closer look at the bergs, but they were farther away than we realized. Everything in the Arctic is bigger or farther away than it appears. At the edge of town, where several dozen sled dogs were tethered to stakes and left to entertain themselves, two decomposing whales—they looked like belugas, but in their condition, without heads or tails, they could have been narwhals—were being fed upon by gulls and ravens, though I suspect they were put there for the dogs to eat. To the uninitiated, they were pretty gross—innards scattered about, coagulated blood, and rotting flesh. Later, we almost stumbled upon the carcass of a dead ringed seal, its fat body nearly concealing the presence of its disproportionately tiny head.

Our boat ride to Koluktoo Bay, our ultimate destination, was postponed and postponed again, partly due to the weather and partly due to the decidedly casual rhythm of daily activities in the High Arctic. As one who likes to keep to a tight schedule, I found this to be frustrating. But after a day-and-a-half of waiting and wandering, we pulled on bright orange survival suits as ten days of provisions were loaded onto a small boat, and headed west. Our guide was Namen Inuarak, a soft-spoken native whose father, brother, and girlfriend provided logistical support as Renay and I and three other adventurers sought close encounters with narwhals. The waves were high, the seas were rough, and the small boat went fast, causing me a little concern at first, but soon I was resigned to a wet and arduous trip. Numerous icebergs of all sizes, shapes, and colors came into view, and at one we slowed to a crawl as fulmars swarmed the area and several seals appeared—our first two harp seals, their dark faces and light shoulders giving away their identities, and several ringed seals. They popped their heads up briefly and watched us with their big dark eyes, but they were hard to see in the rough seas and soon disappeared. After two hours of uncomfortable travel, we turned south past Ragged Island and into Milne Inlet, which leads to Koluktoo Bay. The water was finally comfortably calm, though evidence of birds and other wildlife had disappeared.

An hour from our remote campsite, we rounded a steep, rocky point known as Bruce Head at the entrance to the bay, where we were surprised to see a team of narwhal researchers camped precariously high on the cliff. We waved to the biologists and pulled into a tiny, placid cove a hundred yards beyond their perch to pick up some equipment that Namen had stored there—a kayak, some camping gear, and a steel drum of gasoline. We were watching a couple of seals swimming at the surface nearby, when Renay pointed out something near the opposite shoreline, perhaps a half mile away. Our first narwhals! The animals were almost invisible to the naked eye, and had we not been expecting them, they wouldn't have been identifiable to me as narwhals, even through binoculars—just flat backs that remained at the surface briefly, arched slightly, and then disappeared. No tails visible, no noticeable blows, and not a hint of tusk. As we watched, I could occasionally make out a pattern on their backs—something vaguely black and gray, but nothing more. We scanned the water with enthusiasm for fifteen minutes and ultimately saw three different groups of two or three narwhals. The only distinctive feature was the absence of a dorsal fin, which biologists speculate is an adaptation for living under the ice. At least the trip wouldn't be a bust, I thought, but it was surely an unsatisfying beginning.

When we finally arrived at our campsite, the weather was ominous. The winds picked up dramatically and it began to rain while we unloaded the gear, which made setting up the tents difficult and made me dread what would likely be an uncomfortable night of sleeping poorly on the soggy ground. But it wasn't. It had been a long day, and after a late-night dinner of chili, we climbed into our warm sleeping bags. Despite the worrisome and noisy winds and increasingly heavy rains, my tent stayed dry and sleep came easily.

× × ×

My enthusiasm for the upcoming days of narwhal watching could not have been greater. I thought of the early explorers, who shared exciting tales of bizarre horned serpents living among the seals and whales of the Far North. Prior to our trip, I had spent several days in Massachusetts at the research library affiliated with the New Bedford Whaling Museum, the oldest and largest museum of its kind in the United States, located in the whaling industry's most celebrated port. There, I uncovered some of the earliest known writings about narwhals. Each ancient, leather-bound book I opened showed increasingly unusual illustrations of what early explorers believed the narwhal looked like. The oldest volumes, like those by Swedish writer Olaus Magnus in 1555 and naturalist Konrad Gesner in 1558, written entirely in Latin, included prints illustrating what they identified as Monocerote or Monocerotis, an early variation on today's taxonomical identifier, *Monodon monoceros*. Next to an illustration of a sawfish—which several authors believed was closely related to narwhals—Magnus showed a fish-like creature with a short, thick horn emerging from its forehead. He described it as "a sea-monster that has in its brow a very large horn wherewith it can pierce and wreck vessels and destroy many men." Gesner's narwhal, depicted with just its head emerging from the water, looked more like an alligator, with a long snout and abundant teeth, large dog-like eyes and, like Magnus's creature, a horn sticking straight up from the top of its head.

Surgeon Ambrose Paré, in 1582, writing in French about unicorns, described a sea unicorn "with a horn on his forehead like a saw, three feet long and a half, and four inches wide, with its two peaks very acute." The accompanying lithographic illustration featured a monstrous creature labeled "Vletif" with a long, scaly body, a mermaid tail, two pairs of pectoral fins that looked to me like dragon wings, a wolf-like face and teeth, and a horn like the rostrum of a sawfish rising from the top of its head. It exhibited an angry look as hunters poked at it from land while whaling ships sailed by in the distance. This same illustration was reproduced by Italian naturalist Ulisse Aldrovandi twenty-six years later in his compendium on fish. The book also included an accurate representation of a narwhal tusk, but the author didn't appear to make the connection between the tusk and the real animal.

In 1655, Danish physician Olaus Wörm had similar difficulty. His book *Museum Wormianum* includes beautiful and accurate illustrations of narwhal skulls, obviously drawn with a specimen in hand, but the depiction of the narwhal itself looks more like a dolphin or a tuna, with two dorsal fins, one long ventral fin, a narrow fish-like tail, and a spiral horn emerging from its nose. Similarly, theologian Isaac de La Peyrère in *Relation du Groenland* (1663) features three views of a narwhal skull alongside an illustration of a narwhal looking more like a fish than a whale but with a bear-like snout and its tusk growing from its nose. It is clear that these authors had access to narwhal skulls and tusks from which to create accurate drawings, but they had not been able to study complete narwhal specimens and were instead left to interpret brief sightings of the animals in the water.

In 1700, ship's surgeon Pierre Martin de la Martiniére published observations from an Arctic expedition in *A New Voyage to the North*, which includes a depiction of a narwhal hunt that the Whaling Museum's librarian said "has no precedent." Yet it is the image of the narwhal that was most disturbing to me: a scaly body, hawk-like beak and eyes, and a spiral tusk emerging from its forehead, with a head the size of a rowboat. Written descriptions of narwhals take on similarly exaggerated characteristics. In *The Entertaining Correspondent, or Curious Relations, Digested into Familiar Letters and Conversations* (1739), G. Smith wrote about "a variety of wonderful creatures, which have been discovered, either in the seas, or rivers," including this rather outrageous depiction of fanciful narwhal relatives:

> In the great Northern Ocean are found a Sort of Whales whose Heads are quadrangular, and are all over full of Prickles; besides which they are furnish'd with two very sharp-pointed Horns, each

of ten or twelve Foot of Length. . . . In the same Northern Quarter of the Ocean, is another Sort of Whales, 300 Foot long: Their Throat is twelve foot wide, and their Teeth stand like those of a Boar: The Hollow of their Eyes will hold twenty Men each: About the Eyes, instead of Hair, they have 150 Horns, each seven Foot long, and of a pretty hard substance. About the Greenland coast are catch'd another Sort of Whales, which are 30, 40 and 60 Foot long: They have a-top of their Head two Pipes, through which they cast the Water to a prodigious Height: Out of their Mouth grows a Horn, five, six, seven and sometimes ten Foot long, which grows fast to the Upper part of their Jaws: They shed those Horns as Children do their Teeth; for it has been found that some of them saw'd asunder near the Root, they have been hollow there, and small Horns have fill'd the Vacancy.

Not long afterward, descriptions and illustrations of narwhals became more true to life. Explorer Henry Ellis's *A Voyage to Hudson's-Bay* (1748) was the first I found to correctly illustrate the tail of a narwhal as a horizontal fluke as opposed to the vertical caudal fin of a fish tail that most earlier illustrations feature. The animal depicted was appropriately gray in color with slight speckling, and it had a small dorsal ridge rather than a dorsal fin and one pair of small pectoral fins. All in all, it was much more accurate than almost every other previous illustration. Bernard O'Reilly's written description of narwhals in *Greenland, the Adjacent Seas, and the Northwest Passage to the Pacific Ocean* (1818) more accurately describes their appearance and behavior, though critics claim most of his book is fiction. He must have had a specimen available to him, as he included such details as its tongue length and placement in the mouth ("very short, immoveable, and placed very far behind"), the size of the throat passage to its stomach ("very small, not three inches over"), and its mouth size ("very small; its greatest expansion being not more than six inches"). O'Reilly wrote:

When a number of these animals are together, they divert themselves in playing, when, their teeth appearing above the water, as if brandished about, have a singular effect; and the clattering noise they produce in this confused gamboling, would lead an inattentive spectator to suppose that some hostile proceeding was going forward, which is by no means the case. This has reference to the pacific habits of the monodon; but certainly such an extraordinary provision of annoyance could not have been dispensed for ornament sake; and though the creature being destitute of teeth in the mouth, and subsisting on mollusca and marine vegetables, seems little calculated for destructive or predatory life; yet this tremendous weapon must render him formidable to every inhabitant of the deep that obtrudes upon his peaceful haunts.

William Henry Dewhurst, a surgeon on an English whaling ship, didn't share O'Reilly's belief in the narwhal's pacific habits. In *The Natural History of the Order of Cetacea and the Oceanic Inhabitants of the Arctic Region* (1834), Dewhurst wrote that the narwhal "is an animal possessing almost colossal strength, inasmuch as it precipitates itself upon every thing giving it the least offense, and furiously rushes against the most trifling obstacle. . . . This giant of the frozen ocean dares every power, braves every danger; and bent upon carnage, he attacks without provocation, combats without rivalry, destroying without necessity; and the only enemy to whom he is occasionally compelled to yield to, is man."

Explorer William Scoresby, in 1820, gave a detailed description of narwhal ecology in *An Account of the Arctic Regions, with a History and Description of the Northern Whale Fishery*, which included this early observation of narwhal communication:

A great many narwhals were often sporting about us, sometimes in bands of fifteen or twenty together: in several of them each animal had a long horn; they were extremely playful, frequently elevating their horns, and crossing them with each other as in fencing. In the sporting of these animals, they frequently emitted a very unusual sound, resembling the gurgling of water in the throat, which it probably was, as it only occurred when they reared their horns, with the front part of the head and mouth out of the water. Several of them followed the ship, and seemed to be attracted by the principle of curiosity, at the sight of so unusual a body. The water being perfectly transparent, they could be seen descending to the keel, and playing about the rudder for a considerable time.

There was still great debate, however, about how many different varieties of narwhals there might be plying the waters of the Arctic Ocean. That disagreement was mostly settled by Robert Hamilton in 1837 when he wrote what was considered the most scientific treatment of the narwhal to date. Writing in *The Naturalist's Library* and referencing the speculation of earlier observers, he noted:

> This long established genus [*Narwhalus*] had been very much the creation of fancy; and as, from the peculiarity of its singular horn, it excited much interest, so the errors connected with it have been very widely spread. Both Lacepede and Desmaret have three species; the first of which, the large headed, was represented as a great oval animal sixty feet long, with a murderous horn of sixteen feet in length, with which, in troops, it waged deadly warfare with the mightiest inhabitants of the ocean, and made them an easy prey. This species, however, is a mere figment, the product of ignorance and exaggeration. The third species, *Narwhalus Andersonii*, does not rest on any more solid foundation, so that the genus really comprehends only one species.

The increasingly accurate and scientific descriptions coming to light in the 1800s didn't stop legendary author Jules Verne from reverting to age-old images of narwhals in his 1870 classic *Twenty Thousand Leagues Under the Sea*. His main character, Captain Nemo, a marine biologist, said: "The ordinary narwhal, or unicorn fish, is a kind of whale which grows to a length of sixty feet. . . . [It] is armed with a kind of ivory sword, or halberd, as some naturalists put it. It is a tusk as hard as steel. Occasionally these tusks are found embedded in the bodies of other kinds of whales, against which the narwhal always wins. Others have been removed, not without difficulty, from the hulls of ships which they had pierced clean through as easily as a drill pierces a barrel."

As I soon found out, the truth about the narwhal's abilities, from its flexible tusk to its extraordinary diving abilities, is far more extraordinary than Verne's fiction.

TWO

WHALE SPOTTING

WITH DEPTHS OF UP TO SIX HUNDRED FEET, KOLUKTOO BAY lies at 72 degrees north latitude, north of the northernmost point of Alaska, 2,200 miles north of New York City, and 1,200 miles from the North Pole, where the sun never sets between late May and early August and never rises from mid-November until the end of January. It's in the northernmost region of Baffin Island, the largest island in Canada and the fifth largest island in the world, which is separated from the world's largest island, Greenland, by the seven-hundred-mile wide Baffin Bay and the Davis Strait. Despite the challenges getting to this remote location, Koluktoo Bay may be the most accessible place in the world to get a close-up look at a narwhal.

When we awoke the next morning, the wind and rain continued to pummel our tents, and our boat was unexpectedly gone. Namen's father had taken it to ferry around the advance team for the Discovery Channel television program *Survivorman*, leaving us to fend for ourselves, uncertain when the boat would return and with no means of communication. A Coast Guard ship was in the bay, and a helicopter was constantly taking off and landing from its bow for hours, probably conducting field exercises, so there was some comfort that we could be rescued if necessary.

Our campsite sat on a windswept beach in a protected cove with an abundance of small flat stones perfect for skipping into the calm waters. As the day wore on and the rain eased to a modest mist, we climbed the hillside to watch for narwhals, but none were visible. Namen eventually joined us and pointed out mountain sorrel, a low-growing wildflower with leaves that are tart but tasty and would make an excellent addition to a salad. Inch-high blueberry bushes, some with unripe berries, were mixed in with a surprising variety of other

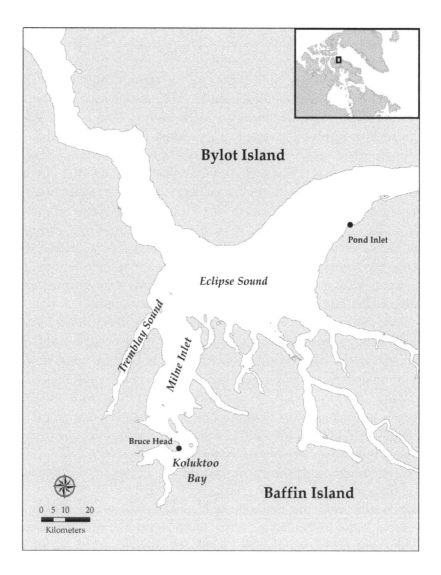

Bylot Island

Pond Inlet

Eclipse Sound

Tremblay Sound

Milne Inlet

Bruce Head

Koluktoo
Bay

Baffin Island

0 5 10   20
Kilometers

tundra flowers, including buttercups, saxifrage, lousewort, poppies, and the distinctive nodding bladder-campion, whose purplish bladders containing its flowers hang down like tiny Japanese lanterns. All have evolved to grow low to the ground to avoid the ravages of the Arctic wind, including the only tree within a thousand miles, the Arctic willow, whose thin trunk runs along the ground like a vine and shoots up leaves extending just three or four inches tall. Numerous small holes and passages beneath the plant matter were evidence of lemmings, and droppings of several varieties indicated that hare, foxes, and caribou might show up at any time. But none did while we were there.

After lunch, Renay and I walked up the opposite hillside and sat in a nook protected from the wind to continue our narwhal watch. We peered through our binoculars at a fish-eating duck called a red-breasted merganser as it rested on the shoreline. It was the third species of bird we had seen that our tattered field guides told us should not be present. I suspect that little data is available from the High Arctic to ensure the accuracy of field guide range maps. After about twenty minutes I saw what appeared to be a whitecap thirty yards off shore in an area of calm water. Thirty seconds later, it turned into a beluga whale, its white body visible even before and well after it surfaced, its large head reaching far out of the water when it arose to breathe. It swam parallel to the shoreline and surfaced several more times, but disappeared before we could alert the rest of our group.

We waited through another day of mediocre weather without a boat until finally Namen's brother arrived with a large wooden canoe that appeared as if it could barely hold us all safely. By then, however, we were anxious to get on the water and would have ridden aboard nearly anything that could float. So at 9:30 the next morning, with calming winds and blue skies, we headed across the bay to where a large group of gulls and fulmars were active, a sign that whales might be in the area. But when we arrived we found no narwhals.

As we continued to slowly traverse the bay, we saw an odd glimmer on the horizon. Through my binoculars, it looked like a rapid series of whale blows reflecting the sunlight. It could have been two whitecaps slapping together, but the water was quite calm and no whitecaps were visible nearby. So even though I didn't actually see a whale—just the blow—I announced they were narwhals. I regretted it almost immediately, not wanting to offend our guides or appear to be a know-it-all when I had no experience in the area and had no reason to be confident in my identification. But when I saw it again, I was even more certain, though I kept my mouth shut. When we approached the area, no whales were to be found, and our group could hardly contain its disappointment.

Namen turned the boat toward Bruce Head, supposedly the best vantage point for watching narwhals, but before we got there, the tusked creatures surrounded us and my pulse quickened as the volume of activity grew. The first animals were somewhat distant and looked like what I imagined to be pilot whales—sleek, black and torpedo-like, but without a dorsal fin. Their blow was puffy, their breathing was audible, and their rounded snouts appeared tuskless. Soon the animals edged closer in groups of two or three, and we were able to make out their markings. Dark whales had slight pale smudges, lighter ones were mostly gray with dalmation-like black and charcoal blotches. We never saw their tails, and given how much of the head was visible without seeing a tusk, I was convinced they were all females. They had long glossy backs when surfacing and, when viewed straight on, some exhibited a dorsal ridge that looked almost like a low elongated fin. In half an hour we must have seen twenty narwhals—never great views, but we were all excited that we finally had some action. Fulmars were everywhere, the sounds of the narwhals blowing and inhaling were boisterous, and the scenery reminded me of an earlier visit to Nome, Alaska, where distant rocky hillsides hid musk ox and grizzly bears and the cerulean sky lapped at the calm waters. Even a few ringed seals put in an appearance to break up the action and confuse us momentarily.

I couldn't have been more pleased. Wildlife encounters are moving experiences for me, leaving me both stimulated and calm and providing an almost spiritual awareness of the natural world. Koluktoo Bay became my cathedral.

While the marine life teemed around us, Namen headed over to Bruce Head where the researchers were positioned and where his brother was camped out waiting for us. He said we would stay there on the rocks for a while, which worried me and the rest of the group since we wanted to return to the narwhal grouping we had just been observing. It didn't help that, when we arrived, Namen's brother said he hadn't seen many narwhals. They told us to be patient, that the steep point of rocks was the best place to see whales.

And it was. But it took some time. I chatted at length with the researchers, who turned out to be two Canadian graduate students I had emailed with eight months previously. As the students went to work high on the cliff, I stared into the open water and the action began. Narwhals started appearing near and far, heading out of the bay and into Milne Inlet. Many were clearly just traveling—swimming in one direction with purpose. They showed long sleek backs, never flashed their tails, and their tusks were never visible. The students said that the wide variety of colors and patterns we observed were an indication of age—old narwhals are pale, young ones are dark. Most of the speckling that we could see was located in the middle of their backs, with little pattern noticeable near the head or tail. Eventually we observed some behaviors that appeared somewhat playful, with some whales actively nudging each other, changing directions, and rolling over, but still they all were moving out of the bay.

While watching a relatively close-in group, I noted a narwhal far in the distance that appeared to lift its tail flukes out of the water, the first time we had seen that behavior. It was too far to see color, but clearly its unusual heart-shaped tail lifted high in the air and then slipped softly beneath the surface. And then another one did it, and another. But never did any of the nearby narwhals raise their flukes. The flurry of activity lasted fifteen minutes, and then Renay and I climbed the steep cliff to the researchers' overlook to chat more with them. The view was breathtaking, reminding me of Barry Lopez's comments about the region: "The beauty here is beauty you feel in your flesh. You feel it physically, and that is why it is sometimes terrifying to approach. Other beauty only takes the heart, not the mind. I lost for long moments my sense of time and purpose as a human being." While few whales remained in the area, the high angle allowed us to see them below the water line, both before and after surfacing to breathe. It even gave me a good view of a couple of narwhals with tails that were patterned and colored similar to their bodies. But still no tusks.

When we climbed back down the cliff to rejoin our group, several young narwhal hunters toting rifles had arrived, and the rest of our group wanted to leave before the shooting began. I had mixed feelings. The day was young, and I was anxious to continue watching the movement of the narwhals from the comfort of land. Our high perch provided an unequaled view, but the last thing I wanted was to watch a narwhal suffer a painful death, particularly since we paid an exorbitant sum to enjoy the majesty of the animals and their quiet surroundings. So, reluctantly, we loaded up the canoe and left.

× × ×

Throughout our first day of narwhal watching, one constant was a large iceberg opposite Bruce Head that we used as a guidepost. It may have been that day when I first heard it said that you never forget your first iceberg, and many months later I couldn't agree more. When I talk about my Arctic adventures, it is the icebergs that I speak about with the greatest reverence and awe. And while the aerial views from the plane and the bergs we saw in Eclipse Bay on our walk that first day in Pond Inlet were impressive, it is the iceberg at the entrance to Koluk-too Bay that I still dream about.

The mass of ice was about half a city block in size, perhaps fifty feet tall, with a tunnel through its middle of a size that a dinosaur could traverse. The tunnel was an aquamarine ice cave that sparkled a cool Listerine blue from sunlight shining through from above. The top and sides were startlingly bright white, but when the sun was hidden, the exterior looked like it had a matte finish, a result of accumulated dust and dirt. Rivers of meltwater poured from the top and sides, carving jagged canals into the hard surface and creating a heavy downpour around its edges. I had never seen anything like it, and its immensity was almost frightening because I knew it could shift or crumble at any time.

What was particularly notable was the quantity of wildlife that seemed to be attracted to the icebergs in the area. Seals were almost always found in close proximity to bergs, though seldom on them, and gulls, fulmars and other birds often circled the icy chunks as well. Namen said that some Inuit fishermen claim that the fishing is often best in the immediate vicinity of icebergs. Recent investigations by scientists are beginning to shed light on what it is about icebergs that is attractive to wildlife.

Ken Smith, senior scientist at the Monterey Bay Aquarium Research Institute, told me that "free-drifting icebergs are hot spots of chemical and biological enrichment." His studies in the Weddell Sea off Antarctica in 2005 and 2008 found high concentrations of chlorophyll, krill, and seabirds surrounding the icebergs he sampled, extending out to about two miles around each berg. While he wouldn't speculate about whether the same results would be found around Arctic icebergs, other scientists say that the evidence suggests that it very well may be the case.

During their first expedition, Smith and his research team deployed a series of oceanographic instruments, trawl nets, and video-equipped remotely operated vehicles in a spiral track around a small iceberg with a surface area of 0.12 square kilometers as well as around another much larger iceberg. Their results revealed how important icebergs are to the marine environment in the Southern Ocean.

Perhaps the scientists' most significant finding was that, as the icebergs melt, they continually release accumulated dust, debris, pulverized rocks, and other terrestrial materials they had scraped from the soil when they were part of glaciers moving across land. When those materials are released into the water, they act as ocean fertilizer to nourish phytoplankton and other marine organisms upon which larger species—like jellyfish, worms, and Antarctic krill—feed. Those species are in turn fed upon by several types of fish, seabirds, and marine mammals.

"These icebergs can be compared to estuaries that supply surrounding coastal regions with nutrients," Smith and his team wrote in the journal *Science* in 2007. Oceanographer Greg Stone had a more hands-on approach to studying the productivity around icebergs. A vice president at Conservation International, one of the world's largest biodiversity conservation organizations, he led a team of scuba divers and scientists on a rare and dangerous expedition to study iceberg ecology around B-15, the largest iceberg ever known, which calved from the Ross Ice Shelf in Antarctica in 2000. Nearly the size of Connecticut and containing enough fresh water to supply the entire U.S. for five years, the berg (dubbed Godzilla by scientists and called "unimaginably large" by Stone) had begun to break up by the time Stone's team arrived several months later, and the shifting pack ice made it impossible for them to reach it. But Stone and his colleagues still conducted about ninety dives around icebergs as large as downtown Boston and as small as an average-sized house. What they found was an unexpected bonanza of marine life in and around the ice.

"You get underwater and it's like an ice cathedral and it seems to go down forever. It's blue and beautiful, but also very cold," Stone said, noting that the water temperature the team dove into was several degrees below freezing.

They saw juvenile ice fish living in holes in the ice the size of wine bottles, though they were uncertain whether the fish excavated the holes themselves to hide from predators or simply took advantage of naturally formed burrows. Like Smith's team, they recorded higher concentrations of jellyfish and krill within a mile or so around the icebergs. In an especially dangerous dive, they swam 1,000 feet into a cave inside an iceberg that had been grounded for three to five years and found a spectacular garden of invertebrates—feather duster worms, starfish, colorful sponges, sea cucumbers, and more—living on the seafloor at the farthest reaches of the cave.

According to Stone, most Antarctic icebergs get pushed around continuously by the wind and currents, and their keels scrape along the seafloor and disturb the benthic ecosystem. But because this particular iceberg had been grounded for several years, it served to protect bottom-dwelling creatures from the ravages of moving icebergs and nurtured this underwater garden.

Stone agrees with Smith's conclusion that the big picture of iceberg ecology is the vital role they play in fertilizing the Southern Ocean, thereby creating the largest biomass on Earth. "The icebergs themselves create these fertile conditions in the ocean because they actually have the nutrients stored in them," he explained. "The ice melts, the nutrients enter and fertilize the sea, and it increases productivity around the icebergs. It creates algal blooms that feed the krill that in turn feed the fishes and larger organisms. The whole food chain gets fired up around one of these things."

The ice, he added, is the biological engine that drives the ecosystem, fueled further by the Antarctic's twenty-four-hour summer sunshine. "There's no doubt that if you're an ocean animal, you're going to be attracted to an iceberg because it's generally a better place to be."

If you're not an ocean animal, on the other hand, it's probably best to keep your distance from icebergs, because one thing they are not is safe. Just hours after Stone's dive team emerged from exploring the cave within the grounded iceberg, the entire berg completely crumbled and disintegrated in one giant implosion, leaving in its place what Stone described as "two square miles of shattered crystal." The fissures formed by the melting and refreezing ice throughout the mountainous structure ultimately led to its destruction, just as the Koluktoo Bay iceberg tunnel collapsed shortly after we left. As frightening as it was to Stone and his team of scientists, who decided then and there never to dive in an iceberg cave again, it also may have been the first time anyone witnessed such an event.

x x x

There are plenty of first-time observations still to be made regarding the natural history of narwhals, too, but slowly the whale is revealing itself to the handful of scientists who are studying it. They know, for instance, that narwhals can weigh as much as 3,500 pounds and grow up to eighteen feet long—excluding the tusk—and that calves are about four feet long and 175 pounds at birth, but there is a great deal of uncertainty about their social structure and the composition of pods. They know that narwhals reach sexual maturity at between six and nine years of age and that females deliver young about every third year, but where they mate or give birth is unknown. They know that narwhals typically swim at speeds of three to nine miles per hour and that they occasionally do so upside down, though no one is certain why. They know that narwhals communicate using a wide variety of whistles, moans, and clicks, but it is still a great mystery what they are trying to say or how they use their echolocation to find food. And they know that narwhals can dive more than a mile deep for as long as twenty-five minutes at a time to feed upon Greenland halibut and other bottom-dwelling fish, Arctic cod and squid, but why they feed very little in the summer months is unknown.

One question that is slowly coming into focus is their population size. Pierre Richard, a research scientist with the Canada Department of Fisheries and Oceans, has been studying marine mammal populations in the Canadian Arctic for thirty years, and while he has seen narwhal population estimates change considerably over the years, he's not sure that the actual population numbers have changed much. What has happened instead, he told me, is that the methods for estimating whale numbers have improved, making the estimates more accurate. "In my experience, the more you look the more you find."

Despite the fact that narwhals are listed in the Convention on International Trade in Endangered Species, Richard estimates that there are about 80,000 living in the Canadian Arctic in August. Of that number, he thinks there are about 5,000 in northern Hudson Bay, another 40,000 around Somerset Island, and the remainder on the north and east coast of Baffin Island, with a few scattered sparsely as far south as Cumberland Sound. Having said that, Richard notes that those figures are mean estimates, but there is a wide margin of error because it is impossible to count every animal at the surface everywhere in their range. That's why scientists sample the population and use all sorts of challenging mathematical calculations to infer the total population.

For example, he said, if scientists sample one square kilometer for every ten square kilometers of their range, and if they see one hundred whales in that square kilometer, they could then multiply by ten to get the total estimate within the range. But it's much more complicated than that. That population estimate would only be true if the animals occurred in the same density in every square kilometer of their range, which they don't. "They could be clumped and aggregated," Richard said. "Therefore your estimate is imprecise due to this clumping. So a measure of precision is calculated, and you get the so-called confidence intervals, and these could be wide."

Part of the difficulty of these calculations is that the scientists can only count the animals that are visible at the surface. Since narwhals are deep-diving creatures who spend about two-thirds of their time far beneath the surface, that behavior has to be taken into consideration. And that means that there are probably three times as many narwhals in an area than you see at the surface. "But that is also imprecise," Richard noted, "because in fact there are times that they spend more time at the surface and times less. There is a measure of precision for that [in the calculation], too. That estimate is a bit more precise, but it adds to the imprecision of the final estimate." It means that any estimate of a population size must include a relatively large margin for error, which could be plus or minus fifty percent or more.

Richard wasn't finished with me. In his thick French-Canadian accent, he talked about observer bias and detection probability and other factors that I'm not entirely sure I understood. He said that when conducting aerial surveys of narwhals and bowhead whales, the pilot flies at about 1,000 feet above the water while two sets of observers are positioned on each side of the aircraft and shielded from seeing each other in order to calculate a measure of observer error for each person.

"They look out the window, and when they see a group of narwhals or an individual, they use a declinometer to measure the angle to the whale sighting, they note the species, the number of animals, the composition of the group, and any obvious ways of differentiating age or sex, and that goes on along a linear transect that is a sample of the area," Richard explained. "You do many of these transects and then extrapolate to the whole area that you census." Aerial surveyors use a hand-held declinometer, a magnetic instrument somewhat like a compass, to model the detection of animals from the aircraft. The farther away the whales are the less likely they will be detected by the observer, and the closer they are the more likely they are to be detected, unless they are immediately beneath the plane, where they are often missed entirely.

It all sounds very scientific and technical and, frankly, not very fun, even though I have long imagined that flying low over the ocean to observe marine mammals has to be an amazing experience. Maybe a veteran whale counter like Richard, who has flown dozens of aerial surveys over a lengthy career and who was preparing to retire, can't generate the enthusiasm for the job that a rookie can. So I talked to Joe Guay, who said he "fell into a gig" with Richard and the Department of Fisheries and Oceans in 2009 while still a student at the University of Manitoba. He started out sorting photographs from aerial surveys, making maps of the results, and spending long hours in an office processing data. But when he got his first chance to fly in a survey plane, he jumped at it.

"It's awesome," he said, flashing the big smile that I'm sure I would have had on my first flight, too. "The cool thing is that we needed a port in the bottom of the plane [to mount a camera], so you're literally flying and there's a hole with a thin sheet of Plexiglass below your feet and 1,000 feet down is the Arctic Ocean. It was just as interesting to look out the side of the plane, where I'd see glaciers, oceans, icebergs, and that, and then I could just bend over this three-foot diameter hole to look straight down into the water. It was really cool looking out that thing." That was precisely the reaction I was hoping for.

Guay said that because surveys can only be conducted in good weather, they spent long days in the air on the infrequent days when the weather cooperated. Stoked up on coffee, he said the experience "was exhilarating, but after staring at the ocean for ten hours, it gets a little straining."

His favorite moment did not involve narwhals, however, but the breaching of bowhead whales, the monstrous baleen whales of the north with the distinctive white chin patch. "That was the coolest thing I saw by far," he told me. "They look dark brown or black from above, but then you see a big white spot as it erupts from the water, and then you see the animal in the air. Pierre had spotted the whales, and the pilot was diving all over looking for them. So not only are we looking at these bowheads breaching, but the plane is going everywhere tilting on its sides so we could see them better."

I'm pretty sure that would have been more than my stomach could take, but there is no doubt that I would have done it anyway.

THREE

A SYMPHONY OF MOOS

IT TOOK MONTHS OF SITTING AT HIS DESK IN WINNIPEG crunching numbers from censuses from 2002 to 2004, correcting for all the potential biases and errors and other factors, before biologist Pierre Richard finally felt comfortable in his estimate of 80,000 narwhals in Canada. That estimate includes a margin of error that suggests that the population could be as low as 25,000 to 30,000 animals and as high as several hundred thousand.

"There is a cost to all these corrections that we do, which means our estimates are less precise than when you didn't correct for them," said Richard. "It's the difference between precision and accuracy." He said that when shooting a shotgun, the shooter may strike the center of the target, but the pattern of pellets is also wide, which makes the shooter accurate but not precise. If the pattern was tight but it was off the mark, then they would be precise but not accurate. "I think [earlier estimates were] precise but not accurate," said Richard. "They were too low on average. But I think that our estimates are more accurate but less precise. We can't escape that. It's the nature of the fact that animals are not behaving nicely for us to get neat, tidy little estimates."

Not everyone agrees with Richard's estimates, however. James K. Finley, an independent Canadian biologist who has studied Arctic marine mammals for thirty years and who is often at odds with official estimates of whale numbers, is highly skeptical that there are 80,000— or even 50,000—narwhals living in the Canadian archipelago. He has flown tens of thousands of kilometers of surveys over narwhal summering areas and has spent many seasons on the ice and on coastal headlands, and as a result he says that he has "a fairly good feel of what abundance means."

He told me that Richard's estimates just "don't seem feasible" to him, because large carnivores like polar bears and lions and narwhals are typically sparsely distributed, especially when they live in the Arctic, which he calls a barren ecosystem that is unable to support large numbers of such animals. While Finley admits that he doesn't have any firm data to back up his skepticism, he added that in his aerial surveys of belugas, which, like narwhals, are highly visible from the air, he developed "a good sense for what 500, 1,000 or 5,000 [whales] looks like. These species both occupy similar habitats, and I did not get the sense that narwhals were two, three, or four times proportionately more abundant than belugas," which would be the case if Richard's narwhal numbers are accurate.

Still, Finley said he would be happy to be proven wrong, and Richard believes he is. Since Richard said his recent census is the first comprehensive narwhal population assessment, there isn't enough historic data range-wide to identify whether population numbers are trending upward or downward. Part of the difficulty of establishing trends is the fact that narwhals move around a great deal, and they don't necessarily return year after year to the same area, so just because a site had large numbers of narwhals one year and fewer the next does not mean the population declined. Most of the narwhals that summer in the Canadian archipelago spent the winter in the pack ice of Baffin Bay and the Davis Strait. Richard said that some narwhals appear to have some fidelity to a particular summering area, but that isn't true of all individuals.

There do appear to be some sub-populations, however, that are segregated. Hudson Bay narwhals, for instance, do not winter in either Baffin Bay or the Davis Strait, preferring to remain in northern Hudson Bay year-round or moving southeastward to winter in Hudson Strait instead, well south of the wintering range of any other narwhal populations. As a result, the whales are beginning to show some genetic distinctiveness that separates them from other Canadian narwhals. The contaminants in their system are also dissimilar to those of other narwhals. Another study of mitochondrial DNA found differences between narwhals in East Greenland and West Greenland but low genetic diversity among narwhals from eastern Canada and western Greenland, suggesting that the latter populations interbreed freely on their wintering grounds.

× × ×

My group of adventurers didn't have to go far to put the narwhal hunters at Bruce Head out of our minds. We traveled just a half mile back to where we had seen the whales earlier, and it didn't appear that the animals had left during our absence. As narwhals surfaced around us, one member of our group pulled out her hydrophones, dropped the sensor into the water, and let us each listen in on the ruckus the animals made beneath the surface. By the time the headphones were passed to me, the smiles on the faces of those who had listened ahead of me signaled that I was about to be astonished. I'm well versed in the amazing diversity of bird calls in nature, and I've heard plenty of recordings of humpback whale songs, but I could not imagine the sound of a narwhal. With headphones on and all above-water noise blocked out, I heard the riotous barnyard sounds made by the narwhals—an entertaining symphony of moos, grunts, creaky floorboards, pops, and clicks. The waters were filled with the cacophony, none of which could be heard above the water. When my turn came to listen for a second time, I made a point of describing to myself each of the sounds that I heard as best as I could. There were sheep-like baahs, Flipper-like clicks, and very distinct and classic sounding moos, along with fake moos like the imitations my father used to make when I was a child. We couldn't help but laugh at the narwhals' bizarre vocalizations.

The sky was nearly flawless that morning, and as the day progressed, it evolved into a beautiful variety of soft wispy clouds, egg carton patterns, fluffy cotton, and threatening gray thunderheads. While we continued to observe the narwhals, an abundance of birds moved in and circled the immediate area. Most were crow-sized fulmars, atop whose dirty yellowish beaks lay a tube-like nostril used to eject excess salt. They appeared in a wide variety of plumages—from dark and smoky to a white-and-gray gull imitation—casually gliding in arcs above the boat and then darting toward us at eye level. Carl Safina wrote in *Eye of the Albatross* that fulmars lack the ease of albatrosses in flight. They "pump hard, as though flying is work for them. They seem to have a dour, businesslike approach to life. If Charles Dickens had written about seabirds, fulmars would have populated the uncouth lower classes of his novels." The seabirds sat on the water beside us in groups, hoping we would toss them a free meal. Then they loudly pitter-pattered with their feet across the water as they attempted to take off, alternately soaring and flapping with determination. Scattered glaucous and Thayer's gulls occasionally joined them, and a solitary long-tailed jaeger sat in the water momentarily, too, something I had never seen from a bird most often observed far out to sea where it steals food from other birds in what is known as kleptoparasitism. As the boat glided toward the jaeger, the bird gracefully lifted off, its long tail streamers undulating with each wing beat until it reached cruising speed.

We sat silently when narwhal activity slowed, watching and listening to the sounds of the Arctic. The soft sloping hillsides were surprisingly green with tundra vegetation and contrasted with the rocky cliffs etched and cracked with crevices from icy run-off. Distant glaciers continued their imperceptibly slow recession. With the boat quietly drifting, it was easy to tell when the narwhals returned. Their breathing when they surfaced was like thunder, first an exhale, then a slightly higher toned inhale. Once again we found ourselves in the middle of a circle of narwhal activity, most of it on the perimeter perhaps 200 yards away. To the left, three narwhals were traveling north; ahead three more circled each other; and to the right, two played, pushing against each other, then slapping their tails with a splash and diving deep. There was so much activity that we repeatedly spun in our seats to get the best views. When three narwhals surfaced simultaneously less than fifty yards to our left and appeared headed in our direction, we were startled. We held our breaths and waited for them to resurface even closer, but they were gone.

Moments later Namen's girlfriend Pamela excitedly called out "Tusk!" and we turned in unison to see a group of at least six narwhals traveling in a line. The lead animals were quite dark, the next two looked very small—likely yearlings—and the next ones were pale and distinctly patterned. We stared hard through our binoculars. The group was moving fast, and for a moment I thought I saw the second animal show a bit of the base of its tusk, but I couldn't be certain. The next time they surfaced, the tusk was clearly visible, angled slightly downward, mostly pale creamy white but darker near the base. At the same moment, Renay affirmed, "Got it!" But one couple, Walt and Ellen, missed it. This was their second trip to see narwhals, and while they saw plenty of them on their first trip, they never saw a tusk. The whales surfaced again, and again the tusk of at least one narwhal was visible. Walt hesitantly said he saw it, and the next time the whales surfaced, we all saw at least one tusk clearly, if only briefly.

Before we got a chance to process the latest sightings and celebrate seeing our first narwhal tusks, we were surrounded once again, with four narwhals traveling behind us, a solo—perhaps our first loner—slowly arching toward us from the right, and several groups of three or four animals far ahead. We heard the odd whooshes of air, like a breeze through a tunnel, that signified more whales in the distance, but they never came into view before it was time to head back to camp.

×   ×   ×

Pierre Richard grew up in Quebec City and spent most of his summers on the shore of the St. Lawrence River, where he quickly became familiar with a variety of marine mammals, including beluga whales. As a graduate student he wrote the marine mammal section of a field guide to the mammals of eastern Canada, and later he got his first job for the Department of Fisheries and Oceans working on the management of Arctic marine mammals.

"At the time, management biologists in other regions were mostly desk biologists who did analyses and made recommendations to fishermen, but in our region we were in a unique position where I was working on marine mammals for which there was very little information from the Arctic," Richard said. "In those days it was not so much a concern about managing populations as it was just wanting to document what was there." Thirty years later he is much more involved in conducting research to provide a scientific basis for wildlife management and habitat impacts. "They don't teach this at university," he said. "You have to learn by doing. Working in the Arctic is extremely difficult—the enormity of the range of the animals, the difficulty reaching these areas due to Arctic conditions. There is a body of literature on wildlife biology management, but it has to be adapted to Arctic conditions, and that has been my working experience."

Part of that experience was time he spent studying narwhal diving behavior in order to know how much time they spent at the surface where they were visible during aerial surveys. With the help of Inuit hunters, Richard's colleague, biologist Jack Orr, developed a system for capturing narwhals using a large mesh net that Orr first saw used by Danish biologists.

"We basically put gangs of nets perpendicular to the shore in areas where the Inuit said narwhals come close to the shore, a point or some promontory or beach area that they are known to go by," Richard explained, describing work he conducted in northern Hudson Bay. "The nets are not detected by the narwhals sometimes until they hit them, then they get tangled up in them, and immediately the crew put inflatables in the water and lift the net to make sure the narwhal doesn't drown." Once a narwhal was captured, Richard inserted nylon pins in the animal's dorsal ridge, through the skin and fat and out through the layer of skin on the other side. Onto those pins he attached an instrument that rides on top of the dorsal ridge. "In the old days we had instruments that were fairly large and didn't last very long, and slowly the pins would migrate out of the ridge," he said. "Now the instrument is about the size of a matchbox."

That matchbox contains a device designed by a company called Wildlife Computers of Redmond, Washington, that measures time, depth, and location on a small microcomputer. When the whale is at the surface and the antennae on the instrument sticks out of the water, it transmits the data via the ARGOS (Advanced Research and Global Observation Satellite) system to a secure website. The data records, for instance, how many dives a narwhal makes, how long the dives are, how deep they go, and the proportion of time they spend at different depths.

"We learned that narwhals dive very deep," said Richard. "They consistently went beyond the depth limit of our earlier tags. In recent years we've found animals diving in excess of 1,500 meters repeatedly. . . . Because they feed mostly on flatfish and other benthic fish, they are essentially foraging at the bottom most of the time. The rest of the water column is just transit."

For narwhals that spend most of the year in deep water, the transit from the surface to the bottom to feed and back to the surface again is a thirty-minute round trip. During most of the year, they make that round trip ten to twenty times per day. Exactly what they do down there and how they feed is anyone's guess. Some biologists have speculated that the narwhals use their tusk to dig around in the seafloor, but most think the tusk plays no role in feeding, especially when considering that mucking up the bottom wouldn't help much in catching halibut. It would make more sense that the tusk would actually get in the way of effective feeding, like wearing a Halloween mask at the dinner table. While there has been little actual research on the purpose or function of the narwhal's tusk, the question has led to a great deal of conjecture and supposition and more than a little controversy.

FOUR

THE INSIDE-OUT TOOTH

BAFFIN ISLAND IS PART OF CANADA'S NEWEST TERRITORY, Nunavut, which was established in 1999, after decades of negotiations and six years after the signing of the Nunavut Land Claims Agreement. The agreement made the Inuit people of Nunavut the largest private landowners in North America, with outright ownership of an area nearly as large as California, though its total population is little more than 30,000. More than 85 percent of the residents of the territory are Inuit, whose primary language is Inuktitut, which was exclusively an oral language until a written alphabet was adapted from a system created by Anglican missionaries for the Cree in the early 1800s.

While the recorded history of Nunavut only goes back as far as 1576, when British explorer Martin Frobisher sought the Northwest Passage and discovered what he thought was gold in what became known as Frobisher Bay in southeastern Baffin Island, historians believe that there have been continuous human populations in the region for more than 4,000 years. The pre-Inuit members of the Dorset culture living along the east coast of the island may have at one time had contact with Norse sailors, as the sagas of early Viking voyages refer to three lands in the North Atlantic coast visited by Leif Eriksson around 1000 A.D. Historians believe that one of those lands, called Helluland or Land of Flat Stones, was Baffin Island. Almost every beach I visited on the island was littered with flat stones.

By the time Frobisher arrived, the Inuit had driven off the Dorsets and the new landholders' first contact with Europeans was a hostile one, with both Frobisher's crew and the Inuit taking hostages. Other explorers to the region soon followed, but as Kenn Harper wrote in his history of the region, "Baffin Island itself was, at best, only a landmark and, at worst, an obstacle in the path of those searching for a North-west Passage. Its coastline remained largely unexplored."

The region was somewhat ignored for the next two hundred years, until a reinvigorated search for the Northwest Passage began in the early nineteenth century with explorations around the Baffin coast led by John Ross, William Parry, and others. After John Franklin sailed into Lancaster Sound in 1845 in search of the Passage, his large expedition team was never heard from again, though his disappearance ushered in another new era of Arctic exploration, most of which bypassed Baffin Island entirely.

British whalers began hunting bowhead whales off northern Baffin Island beginning in 1817, and when whalers started wintering in Cumberland Sound at the south of the Island in the 1850s, companies established shore stations there. When the whale stocks were depleted by the turn of the century, the stations were converted to trading companies, and mining operations soon opened, including an expedition to Pond Inlet in search of gold. Soon ethnographers arrived to study the native people, among them geologist and filmmaker Robert Flaherty, whose 1922 silent documentary film *Nanook of the North* became an early classic.

Perhaps the saddest chapter in the region's history occurred in the early 1950s, when the Canadian government relocated eighty-seven Inuit from northern Quebec to establish new communities in the extreme far north to further exert sovereignty over the region. While they were promised better living and hunting opportunities and told they could return in a year, these promises were not honored by the government. Left there in harsh conditions without adequate food or supplies and unfamiliar with the environment and wildlife of the region, they suffered great hardships. Although the families eventually learned to survive in the unforgivable landscape, a government investigation in 1993 concluded that the relocation was "one of the worst human rights violations in the history of Canada." The families were offered $10 million in reparations and the opportunity to move back to Quebec, but never received a formal apology from the government.

When Nunavut was established and the Inuit finally gained control of their communities and their future, many of the settlements in the territory reverted back to their Inuktitut names, including its capital Iqaluit, which was formerly called Frobisher Bay. While Pond Inlet has officially maintained its English name, the locals call it by its traditional name, Mittimatalik.

× × ×

Not far away in Koluktoo Bay, it was 11 p.m. and I was still relishing the day, enjoying the glow of an unexpectedly stunning series of narwhal sightings that followed three frustrating days of huddling in our tents. It was officially the last day of twenty-four-hour daylight for that latitude, and while I knew it would still be bright enough to read by all night long for several more weeks, I didn't want the day to end. Everyone else had retired to their tents, but I stood alone in the quiet cove, the water almost still for the first time that week, disturbed only by ripples caused by two frolicking ringed seals playing a silent game of peek-a-boo in the deep waters, and a few unruly Arctic char splashing in the shallows. The sun made an orange glow around distant graying clouds well above the horizon, while blue sky and fair weather clouds danced overhead. The water was a clear blue-green like that of tropical beaches, turning dark and haunting as a cloud crossed the sun. The reverberations of a nearby trickling stream competed for my attention with the sound of my pulse. With the still-bright sky reflecting in the waters of the bay, the tundra hillsides—home to ptarmigan and Arctic hare—a pale green, and the rocky cliffs standing strong and defiant against the crushing blows of extreme weather, I thought that it might have been the most beautiful place on Earth. Despite a few agonizing days, I could finally tell myself that it was good to be there.

On my last glance around the cove before turning in, I heard the unmistakable sound of a surfacing narwhal somewhere in the distance. Two whales entered the cove about a half mile away, marking the first time since we arrived that we could see narwhals from our campsite. I figured it was a good sign for what was to come the next day. What I didn't realize was how quickly the next day was going to come.

As Renay and I finished a crossword puzzle in our tent and prepared to sleep, Renay said she thought she heard a narwhal exhale nearby. She peeked out of the tent and saw an enormous tusk emerging straight up from the water just fifty feet from our encampment. Softly but sternly, she called for me to come immediately, grabbed the camera and binoculars, and woke the rest of the group. We tentatively emerged wearing nothing but our thermal underwear, but our self-consciousness disappeared as we focused on the wildlife encounter of a lifetime.

At first all we saw was a long narwhal tusk reaching toward the sky. It was distinctly spiraled over its entire length, mostly dark gray with pale cream at the tip. The animal began waving it back and forth, sometimes jabbing it high into the air, other times lowering it at an angle and sweeping it across the water, then submerging it completely only to extend it upward again at a different angle. Yet the body of the narwhal was still not visible, not even a slight hint of its head. I assumed—incorrectly, as I would learn later—that it was feeding, as it was right where the char had been jumping earlier. After a minute or so, the whale lowered its tusk, then surfaced, its long back exposed to the air, and slowly swam perpendicular to our beach. It was an unexpectedly dark narwhal, given the length of its tusk, with light gray splotches across its back.

As it disappeared—followed closely by a glaucous gull—several other narwhals entered the cove from the north. One slowly headed in our direction, and many more were made visible by the midnight light reflecting off their wakes, creating softly illuminated lines tracing their paths. We could see six or seven distinct trails in the water and hear the animals breathe, even though we couldn't actually see the narwhals themselves. The narwhal that came our way approached the same area as the previous one, stopped in exactly the same spot and behaved in the same way. His body sank beneath the surface, and then his tusk emerged straight upward, waved back and forth, dropped, then jabbed upward at varying angles. That tusk was probably longer than the first—six feet maybe—but similarly colored. And then he jabbed it so high upward that he lifted his head straight up out of the water, showing the odd jaw line where the tusk emerged from his face—lower than I would have guessed, which may explain why we rarely see the tusk when they swim. Three times he raised his head and tusk to the sky, the last time followed by a gull swooping in to grab something at the surface. Then, to our amazement and surprise, a second, smaller narwhal approached and raised its two-foot tusk beside the other. The two clashed briefly before rolling over each other like they were wrestling. It appeared to be a playful and friendly encounter, not aggressive. The small tusk was nearly all white and stood out against the dark, thick tusk of the adult. After less than a minute, both animals lowered their tusks beneath the surface and swam away side-by-side.

By then there was so much activity in the cove that I didn't know which direction to look. The glistening wakes of other narwhals could still be seen in the distance, each swimming independently and quite slowly. We kept watch for an hour, listening to their breathing, noting each blow, and even seeing a few more tusk displays—probably the same animals, we guessed—far out in the bay. When the nearby activity decreased, we returned to our tents. It was one o'clock in the morning, we had been standing in nothing but our long underwear for more than an hour in the High Arctic, shivering in bare feet, unwilling to take thirty seconds to retrieve our shoes and a shirt from our tent for fear of missing something extraordinary. It was an absolutely spectacular experience, and we were speechless. Later, as I lay in my tent taking notes, we continued to be serenaded by the breathing of surfacing narwhals—some hoarse, some like wind tunnels, others low and hooting and cooing or croaking.

It took a long time for my pulse to slow down enough to fall asleep.

× × ×

Back in Pond Inlet at the co-op, the lone retail store in town, you can purchase everything you might need for Arctic village life. It's a grocery store, hardware store, gift shop, restaurant, clothing outlet, post office, art gallery, fuel depot, and cable television service all wrapped up in one. Clad in drab metal siding highlighted with graffiti and surrounded by windblown food wrappers, the building and its rutted parking lot are reminiscent of Main Street USA, but with all-terrain vehicles instead of SUVs. On its front steps teenagers trade insults with their friends, neighbors chat about the weather and the day's hunt, and inside tourists and villagers conduct their business. It's also the place to go if you happen to be in the market for a narwhal tusk.

When Renay and I visited, we saw seven tusks of varying lengths hidden behind a door to the store manager's office, too valuable to have on display (though how anyone would steal a seven-foot spike of ivory without being seen I can only imagine). Each tusk was tagged to document that the narwhal was captured legally, with the name of the hunter, the tusk's length, and the date the animal was killed. The tusks were smooth to the touch, despite the distinct spiral. Dirty stains from marine algae filled the scratches and textured valleys while the raised areas and the entire tip were a clean, creamy bone color. We were surprised by their significant heft—perhaps as much as twenty pounds for a long one, which led me to imagine all sorts of neck and vertebrae injuries if humans had to carry such a tusk around in their jaw. We were equally surprised at the long cavity that extended down most of the length of each tusk, like someone had drilled a vertical bore hole into them. Every tusk was priced at more than $1,000, far out of the price range of any of the local residents, though the locals weren't the target market for the unusual teeth.

When we returned to the store eight days later after our visit to Koluktoo Bay, all of the tusks had been sold. In our absence, a cruise ship had stopped at Pond Inlet, and the vacationers onboard—all from Europe and Asia—bought up every tusk. The Marine Mammal Protection Act prohibits anyone from bringing narwhal tusks or other body parts into the United States, though one shop owner in British Columbia I talked to many years ago said she could get around the law and ship a tusk to me if I paid a little extra. I didn't, but I understood the desire to proudly display a six- or eight-foot long tusk on one's living room wall as a remarkable artifact of nature. After all, I have a collection of animal skulls found in my travels or recovered from road kill, strange evidence of my passion for wildlife and my unusual taste in art and collectibles. But in this case, I decided that photographs of narwhals were a better choice, and legal. Yet my first experience seeing and touching a narwhal tusk certainly primed my curiosity about the singular oddity of its tooth.

I called on a dentist—Martin Nweeia—to share with me some of what he has learned about the narwhal's tusk. A Connecticut resident of Assyrian descent, he has great curiosity about the structure of one of the natural world's largest and most unusual teeth. Nweeia maintains a general dental practice in northwest Connecticut while also serving on the faculty of the Harvard University School of Dental Medicine. As a teenager he had a strong interest in both medicine and art—he's a classically trained pianist who has written scores for public television documentaries—so he chose dentistry because he said it is "a highly artistic profession." He noted that dentistry involves three-dimensional design, especially restorative and aesthetic dentistry and facial reconstruction. "The overall aesthetic is seeing how the face and mouth work together," he told me. "The best restorative dentists are the ones who are the best artists."

As a dental student, Nweeia participated in an anthropological study of the Ticuna Indians of the Amazon basin and later studied the skeletons of other South American tribes to look for indicators of disease. "Teeth often act as a permanent record of disease, and when they do, since teeth are the most well-preserved tissue of any mammal, you have the ability to look at diseases of primitive cultures," he said. Later, Nweeia led a study of childhood dental diseases on the Micronesian island of Yap to see what it could tell him about the health of a very remote native population. As he began to give presentations about these studies to scientists and the general public, he often discussed various examples of teeth in nature—elephants, walruses, and eventually narwhals. In 2000, he decided that the narwhal would be the subject of his subsequent investigations, and those studies continue today.

His first narwhal expedition was to Pond Inlet, and he has returned several times to collect tissue and tusk samples from narwhals killed by Inuit hunters—large organs, full skeletons, and half a dozen heads, all frozen and stored at the Bone Cell Biology Lab at Boston Children's Hospital and other research labs around the globe. The analysis conducted on these samples by Nweeia and his collaborators has formed the basis of the most comprehensive understanding of the development and structure of the narwhal tusk yet undertaken.

Adult male narwhal tusks are typically six to nine feet long and horizontally embedded in the left side of the upper jaw, extending forward through the upper lip like a perverse body piercing. A second, parallel tusk, usually no longer than twelve inches, is embedded in the right side of the jaw without erupting through the lip. In only about one out of every 200 narwhals does the second tusk grow to nearly match the first. This extreme dental asymmetry is virtually unmatched in the natural world, with perhaps the only exception being that seen in the fossil record of *Odobenoceptops peruvianus*, a walrus-like cetacean that may be an ancient relative of the narwhal and beluga. The left-handed spiral of the tusk is also virtually unmatched in nature, although some elephants that undergo trauma shortly after birth may later develop spiraled tusks. In rare cases, female narwhals can have a tusk, but it is usually considerably shorter and narrower than that of the male, an unusual example of sexual dimorphism in mammalian teeth.

The narwhal tusk is a spectacular aberration of natural history, with remarkable structural strength that allows it to flex up to a foot in all directions. No wonder it has provoked awe and reverence and speculation and curiosity for centuries. But the whale's bizarre dentition doesn't end there. In the early stages of tooth development in fetal narwhals, there are actually six pairs of teeth—twelve teeth in total—but four of those pairs disappear before the animal reaches adulthood, and the last pair remain impacted in the upper jaw and never develop further. This latter pair even has a nerve canal system that traces back to the nerve canal of the actual tusk, but their development ends there and they seem to serve no known function.

"If I scanned a sub-adult or an adult narwhal, I would not see any evidence of those four other pairs of teeth, but in the fetus you can see them," Nweeia said. "They are tooth buds, they are locations embryologically where a tooth would normally form, but then they disappear. I hesitate to say that they don't form, because it's possible that you could get early formations that then disappear, but why would that be important? You've got a mechanism in a narwhal that tells you that you've got four of these things that were developing, and all of a sudden a switch told them to turn off. If you can find a genetic key as to why that happens, then you can perhaps do the same thing to a group of cancer cells."

Equally strange is the fact that the tooth that eventually becomes a tusk begins its development in the front of the jaw of the fetus, almost directly in front of the vestigial teeth, but then reverses position and migrates to the rear of the jaw before growing forward again by the time the narwhal becomes an adult. "Typically, when you see developing teeth, you see them all lined up. The fact that these teeth had a dynamic relationship wasn't described in the literature. No one would have believed that, including us," Nweeia said with surprise and wonder. "It's not what I would have predicted."

It got him thinking about why, from an evolutionary standpoint, narwhals started out with twelve perfectly functional, capable teeth that could help it chew and ended up with a horizontally impacted tusk that grows eight or nine feet out through its upper lip into the water in such a way as to almost certainly be an impediment when swimming. "Somewhere along the line that was an evolutionary decision," Nweeia said. "And if you watch a killer whale pursue a narwhal, you realize that the tusk is just the worst thing you could possibly have on you. . . . Why would you do this? Nature doesn't allow you to develop something for sexual display that's a complete disadvantage to you. This thing is a complete disadvantage. There's nothing about it—when you look at it and when you watch one of these things swim—that says this helps the whale."

× × ×

When Nweeia sought partners who could help him answer questions about the physical structure of the tusk, he turned to the American Dental Association Foundation's Paffenbarger Research Center in Gaithersburg, Maryland, and its long-time director Fred Eichmiller, who now directs dental science and research for a dental insurance company in Wisconsin. Like Nweeia, Eichmiller has a dental degree, but he combined that with a degree in mechanical engineering in order to conduct research on dental materials. During his twenty years at Paffenbarger, he developed new dental filling materials, studied materials for bone repair and regeneration, and gained wide-ranging experience studying bone morphology and testing bone tissues. He said that many of the techniques he uses to study human teeth and other tissues were successfully applied to the study of the narwhal tusk as well.

Eichmiller and his fellow dental researchers started by closely examining whole tusks using microscopic photography before cutting, sectioning, and polishing tusk samples. He was particularly intrigued by what he called a "wear facet" or concave depression found near the tip of many tusks. So he asked a ceramics engineer who was an expert in fractology—the study of fracture and wear—to examine it, and she came to the conclusion that it was a wear pattern or abrasion from rubbing it on sand, perhaps from jabbing it into the seafloor. He then turned his attention to the spiral nature of the outer surface of the tusk, which he found to consist of two separate twisting structures.

"There's a minor helix, which occurs during the cellular development of the tusk, and it forms small ridges that are a millimeter to maybe two millimeters deep," Eichmiller said. "And then you have a major helix that will have valleys within these smaller valleys that could be several millimeters up to about a centimeter. And they're both on the same twist rate. When the tusk grows, it grows from the base and it actually rotates as it grows. That would make sense for trying to keep a central axis."

It wasn't until Eichmiller began slicing and polishing the tusk that he made the first of several unexpected discoveries about its structure. "The narwhal tooth was actually made inside-out," he said. "The most highly mineralized and hardest tissue was in the center where the pulp is and the softest and most resilient material was around the outside."

By using infrared microscopy, he examined a polished cross section of a tusk and created a tiny map of it to determine the kind of minerals and proteins it was made of. What he found was that, like most mammalian teeth, it was composed of protein collagen and hydroxylapatite mineral, but the relative proportions of each were very different. "Normally on a chewing tooth you have enamel, which is about 95 percent mineral and very little protein, on the outer side, and when you get into the dentin you get tissue that's mostly protein and much less mineral so it has a toughness to it. But the narwhal tooth is inside-out."

He found that the outer layer of the tusk consists mostly of cementum, a material that typically coats the root of human teeth and serves as the glue between the tooth and bone, which, if it gets exposed by receding gums, is quickly worn away. But on the narwhal this cementum provides a thick leathery coating to the outer layer of the tusk. "From an engineering perspective, that makes an awful lot of sense," Eichmiller said. "Trying to make a tusk that long, you need a high degree of toughness because I'm sure it flexes quite a bit as they swim through the water. If it had a hard, mineral-like outer coating, you'd get cracks forming and you'd have flaws that would lead to fracture. By having an outer coating that's tough and resilient, you don't risk forming cracks."

This inside-out structure makes the tusk both strong and flexible. Eichmiller sawed pieces of tusk into small bars and used a mechanical test to bend the bars to their breaking point as a measure of strength. Human tooth dentin has a tensile strength of sixty to eighty megaPascals, he said, and human tooth enamel has a tensile strength of fifteen to twenty MPa. The narwhal tusk that Eichmiller tested reached 160 MPa at its base where naturally occurring stresses would be highest—more than twice the strength of human dentin—and nearly 100 MPa at mid-tusk.

He also conducted a test of "elastic modulus," which is a way of characterizing the stiffness of a material by assessing how much it can flex before permanently bending. According to Eichmiller, this test can either be done by measuring the rate at which a solid bar of tissue bends or by measuring the rate at which a diamond point penetrates the surface of the tissue. Human dentin has an elastic modulus of twenty to twenty-five gigaPascals and human enamel is more than eighty gigaPascals, which is considered very stiff. The narwhal tusk has a modulus of nearly twenty GPa near the tip where there is little flexing, but this value drops to ten GPa near the base and even as low as one GPa on the outer surface near the base of the tusk where stress and flexibility would be greatest. In cross section, the inner part of the tusk nearest the pulp has a modulus of about ten GPa and drops to less than one GPa on the outermost layers. "What it confirmed," Eichmiller explained, "was the inside-out aspect of the tusk, where the highest hardness and least flexibility is down the center and the greatest resilience and highest flexibility were in the outer layers. . . . It's more rubber-like as you move to the outside, and that's the key to being able to have a tusk of that length with that kind of strength."

×　×　×

The only other marine mammal with decidedly iconic teeth, and the one that may be considered even more charismatic than the narwhal, is the walrus, noted for its large size, stiff whiskers, massive external tusks, and brownish-pink skin. Weighing in at upwards of 4,200 pounds and twelve feet in length as adults—about the size of my Honda Civic—walruses occupy a rather narrow ecological niche, requiring ice-free, shallow water containing an abundance of available mollusks in the seafloor to feed upon and ice or land nearby where they can rest and molt. While they can swim at a great rate over tremendous distances, walruses often do most of their traveling by riding floating icebergs.

Five distinct populations of Atlantic walrus are recognized, including an extirpated subgroup that lived around Nova Scotia, Newfoundland, and the Gulf of St. Lawrence and was hunted so aggressively in the seventeenth and eighteenth centuries that it never recovered. The walrus has been designated by the Canadian government as a species of special concern based on its small numbers and declining trend. The population in south and east Hudson Bay numbers in the low hundreds, making it vulnerable to disturbances and increased hunting; the northern Hudson Bay–Davis Strait population may number 6,000, but it is facing unsustainably high mortality from hunting; in Foxe Basin to the east of Baffin Island, walrus numbers are estimated at 5,500; and the best guess for the Baffin Bay population in the High Arctic is about 1,500, which is only a few percent of the number present a century ago. Including animals in Norway, Russia and Greenland, the worldwide population is about 22,000, far below the 200,000 Pacific walruses that live in the waters of Alaska and eastern Russia. Despite the fact that hunting continues to be its greatest vulnerability, there was no management plan in place for the Atlantic walrus when the species was last assessed by Canadian officials in 2006.

Walruses may have been the target of hunters for thousands of years, as Inuit and their predecessors killed them for their meat, blubber, bones, and tusks. Commercial hunting of Pacific walrus continues in Russian waters and is justified by its healthy population, which doubled between 1960 and 1980. While hunting by humans is still believed to be the largest contributor to Atlantic walrus mortality, the animals are also preyed upon by killer whales and polar bears, and aggressive fighting between males during the breeding season also takes a toll.

Clearly, the most distinguishing feature of male and female walruses is their tusks, which sometimes grow to three feet in length and are essentially oversized canine teeth that grow straight downward throughout their lives, unlike the forward-growing spiral tusk of the narwhal. Male walruses use their tusks primarily in territorial battles or to attract females during the breeding season, but they are also used by both sexes to help haul themselves out of the water or to move around on the ice. This latter use is largely why their genus, *Odobenus*, of which they are the only member, means "tooth walker." It is also not unheard of for walruses to use their tusks to puncture inflatable boats when provoked. Scientists previously believed that walruses used their tusks as a tool for digging clams and other mollusks in the sediments, but now they agree that they feed by creating a powerful suction with their cheeks and sucking marine creatures from their shells.

One concern that has recently been raised about both Atlantic and Pacific walruses is their proclivity for fatal stampedes. Due to the warming climate, there are fewer and fewer icebergs for the animals to haul out onto, so they congregate in larger and larger numbers on beaches and available ice floes. The World Wildlife Fund reported seeing 20,000 Pacific walruses on one Russian beach in September 2009 in an area where several thousand walruses were crushed to death in a stampede in 2007 when the animals were disturbed and dashed into the water. That same month, the U.S. Geological Survey reported that 131 walruses died in a stampede near Icy Cape, Alaska. Similarly large numbers were also found on several Alaska beaches in 2010 and 2011. While it is unlikely that so many Atlantic walruses would be killed in a single event, due to the much smaller size of their population, any stampede could be potentially fatal to an entire generation of juvenile walruses hauled out at any given location.

"Stampedes are frequent and stimulated by many types of disturbance," said Brendan Kelly, professor of marine biology at the International Arctic Research Center at the University of Alaska in Fairbanks. "I have seen large numbers stampede, for example, when a sleeping walrus was startled by a gull calling. Typically, when one animal startles and heads for the water, it sets off a chain reaction as others around that animal also startle."

According to Kelly, the long term impact of using shorelines instead of ice to rest on are not easy to predict, but the impact will likely be negative. Walruses and other seals prefer to rest on sea ice because it provides them with an excellent refuge from most predators. They also protect themselves from predation by forming very dense, large herds when out of the water. Presumably, he suggests, the decrease in the likelihood of being preyed upon compensates for the increased risk of being injured or killed in a stampede.

Kelly said that walruses cannot effectively feed in waters much deeper than 300 feet, yet the retreating sea ice has taken the ice cover over water much deeper than that. This is of particular concern because walrus mothers nurse their young for upwards of two years on protective ice floes, from which they could also feed. The large numbers of walruses coming ashore appears to be a response to this "decoupling," as Kelly called it, of the use of the ice platform as both foraging and nursing habitat.

The biologist is quite pessimistic about the future of the Atlantic walrus in the wild, primarily because of the impact of the loss of sea ice. He said that all of the pinnipeds—seals and sea lions, including the walrus—have traded terrestrial mobility for aquatic mobility, even though they use both environments, feeding in the water and giving birth on land. One result of this tradeoff is that they are slow and vulnerable out of the water and can successfully reproduce only where there are refuges from predation. Since only a tiny fraction of the earth's surface consists of predator-free islands, sea ice provides them with a tremendous refuge for raising their young. Unfortunately, summer sea ice is quickly disappearing, leaving walruses especially challenged due to their prolonged nursing period. "If walruses survive the loss of summer sea ice, I suspect it will be at much reduced levels," Kelly said.

× × ×

I first met Martin Nweeia in the Harvard Medical School's New Research Building, a sparkling glass and metal structure filled to the brim with elegant meeting places and the highest of high-tech laboratory spaces. Walking down one hall lined with tall narrow gray lockers, I was reminded of a high school corridor. Through a door off the hall, I could see what looked like a busy hospital ward, smelling of antiseptic and with carts and charts and equipment lining the beige walls. But at the end of the corridor it opened up into a genetics lab like none I had ever seen.

A dozen or more separate-but-connected research bays, brightly lit from floor-to-ceiling windows, were stuffed with custom-made wood cabinetry and counters covered with biotechnology equipment, computers, chemicals, and everything needed for genomics and proteomics research. According to Nweeia, it's the largest laboratory in the world for conducting dental analyses. There was so much equipment, in fact, that there was hardly any available counter space for the scientists to work. Yet it is there that much of Harvard's most impressive DNA analysis is done.

Winston Kuo, who runs the lab, called it a playground for scientists who want to experiment with the newest technologies: Barocyclers for extracting DNA, RNA, proteins and lipids; NanoStrings for placing "barcodes" on a genome so it can be more easily tracked; RT Analyzers, which Kuo called a one-stop shop for processing genetic samples; and a BioNanomatrix for rapidly sequencing genomes for almost pennies. Amazingly, Harvard didn't have to spend a dime on any of these tools, most of which look to the uninitiated like your basic microwave oven. They were all donated by companies that hoped to gain visibility and accolades from being associated with the research conducted there. More often than not, all Kuo needs to do to get a new piece of technology is to tell manufacturers' representatives that the equipment would be used on a high profile project. And while much of the research in the facility provides insight into cancer and other human diseases, the novelty of a narwhal tusk study has attracted the interest of a wide range of excited collaborators.

A dentist by training, Kuo met Nweeia through mutual contacts in the world of dental medicine, and he was excited to tackle one of the most unusual dental research projects he could imagine—a genetic analysis of narwhal tusk development. While his lab is best known for its studies of cancer and microbes in humans, it does its share of animal work, too, though most of that is to better understand human diseases. When Nweeia said he was sending some narwhal samples, Kuo wasn't sure what to expect.

"All of the other organisms we work with are very small. Even the embryo of a narwhal is really big for us," Kuo said laughing. "We had to scale things up a little bit in terms of sectioning and preparing the samples. It takes a bit longer for some of our solutions to penetrate the narwhal. This is all new to us, so we don't know how long it will take compared to mice, which we're accustomed to. We can always generate a lot of mice samples if we need to, but it's not like we can generate a lot of narwhals. So we're very careful."

The objective of the molecular studies is to identify biomarkers that can shed light on the evolutionary development of narwhal teeth. The researchers want to draw a genetic roadmap that will tell them how and why narwhals develop the teeth they do, but they are still looking for the genes that signify the starting line of the process. This kind of analysis has never been done on whales before, so Kuo and his colleagues are proceeding slowly and deliberately and seeking additional experts wherever they can find them.

Using microCT scans and all sorts of genetic analyses, they hope to understand, for instance, the asymmetry of narwhal teeth, why the tusk spirals, why so few females have tusks, and why it is always the left tooth that becomes the tusk while the right one usually remains impacted in the jaw. Ultimately they aim to sequence the genome of the narwhal and create a library of tissue samples—from early embryonic tissues to adult stage tissues—so future researchers have examples to compare against.

"We're defining the process as we go along," said Kuo, who believes that sequencing the narwhal genome will dramatically jumpstart the understanding of the narwhal. "With the map of the genome, you can learn a lot of things. [You can find] genes that are biomarkers for cancer, or you can look for mutations. All of this is in our recipe. This is what we want to be able to do for the narwhal, to be able to understand what is the recipe or the history of tooth development or of the whale itself. . . . I'm not saying we're going to solve diseases for whales, but to be able to use the dictionary of life of the narwhal to understand it, and perhaps correlate that to human disease."

When I expressed my surprise about the notion that narwhal studies could benefit humankind, Nweeia said that the process is already under way.

> One of the things we found early on is that the [narwhal tusk] tissue itself was even harder than some of the enamel component that human teeth have, and yet it had more elasticity to it. It was more flexible," he said. "There's a group in D.C. trying to clone some of the tissue for possible use as restorative material [for human teeth]. Is it possible to use narwhal tissue? In materials science, you're looking for flexibility and hardness, and usually you sacrifice one to gain the other. To have a tissue that is so extreme in both those capabilities is so unusual.

Back at the Paffenbarger Research Center in Maryland, Fred Eichmiller made what may turn out to be one of the most significant discoveries about the structure of the narwhal tusk to date—tiny tubules extending through the entire thickness of the tusk material, starting from the central portion of the tusk where the blood vessels and nerve endings form the tusk's pulp and ending when they break through the tusk's outer layer. Using a high-powered scanning electron microscope, Eichmiller first noticed the tubules on the interior "pulpal surface" of the tusk and recognized them immediately because they are found in all mammalian teeth.

"Tubules typically extend from the pulp through a portion of the dentin, rarely through all of the dentin, and in odd circumstances you might see them gaining access to part of the enamel," explained Nweeia. "But never do they go all the way through like this."

Eichmiller said that the tubules in the narwhal's tusk are about the same size as would normally be found in human teeth—one to two micrometers in diameter—and were found all the way from the base of the tusk to its tip. In fact, he said that everything about the tubules looked almost exactly as they would look in a human tooth from the inside, except that there were fewer of them per square millimeter.

"It was a pretty interesting discovery," he added. "Totally unexpected."

In human teeth, and in every other animal, there is a coating on the outside of the tooth that prevents the tubule from penetrating to the surface. The dentin coats the surface so those tubules never go all the way through. When you have root resorption—when the root breaks down or erodes away—and the tubules get exposed, that's when teeth get sensitive, especially to hot or cold temperatures. It also occurs when a cavity forms and the enamel is worn away. That's when you get the first sensations of pain.

"When we saw that, it didn't make sense," Eichmiller said. "Why would you have open tubules down the entire length?"

It was the first clue to the researchers that the narwhal's tusk might possibly serve as a sensory organ.

~~~ FIVE ~~~

MYTHOLOGY

NUMEROUS THEORIES HAVE BEEN PRESENTED OVER THE CEN-
turies to explain the function or purpose of the narwhal's tusk. Is it a
spear for hunting or a tool for digging, a weapon of defense or aggression, an instrument for breaking ice or sound propagation, or even a
swimming rudder or an organ for breathing? Even Herman Melville, in
his 1851 classic *Moby-Dick*, weighed in on the subject:

> It does not seem to be used like the blade of the sword-fish and bill-fish;
> although some sailors tell me that the Narwhal employs it for a rake
> for turning over the bottom of the sea for food. Charlie Coffin said it
> was used for an ice-piercer; for the Narwhal, rising to the surface of
> the Polar Sea, and finding it sheeted with ice, thrusts his horn up, and
> so breaks through. But you cannot prove either of these surmises to
> be correct. My own opinion is that however this one-sided horn may
> really be used by the Narwhal—however that may be—it would cer-
> tainly be very convenient to him for a folder in reading pamphlets.

All of these, the serious and the grandiose, have been debunked in
favor of evidence pointing to its role as a secondary sexual characteristic to establish social rank in competition for females. No one had
ever even considered the possibility that it could be a sensory organ of
some sort. But that was the immediate conclusion drawn by dentists
Martin Nweeia and Fred Eichmiller. To me, that seemed farfetched. I
had spent enough time with biologists studying dozens of other wild
animals, from sharks and porpoises to bobcats and falcons, to know
not to jump to conclusions based on one untested observation. These
men aren't biologists, but their passion and certainty made it difficult
not to believe them.

"In human teeth, tubules cause sensation. So we figured these things [in the narwhal tusk] must also have the capability of being sensory," Eichmiller said. "We also know from human teeth a lot of the things that can cause sensation—temperature, osmotic gradients—and that led us to think that there might be a lot of things in the ocean environment that could do the same thing."

"There's a fluid inside the tubule," Nweeia noted, "and the cold actually draws that fluid out. At the base of that fluid is a nerve cell, and when that fluid gets pulled, it pulls a bit of that nerve cell, the thing that sends the perception to you that there is cold. That interstitial fluid is almost identical in its components to blood plasma, which is almost identical in its components to sea water. This is a cool mechanism!"

Ever since the tubule discovery was made, Nweeia has been on a crusade to obtain scientific evidence to confirm his hypothesis that the tusk is a sensory organ. He worked with engineers at the Massachusetts Institute of Technology to create sensors, and with oceanographers at the Woods Hole Oceanographic Institute to build a housing to attach the sensors and computer equipment to the tusk of a live narwhal. Most of the work was done in the basement of his dental practice in Connecticut, which he described as being "like a bake shop and we were doing our own recipe." It took him multiple expeditions over four years and the use of hospital grade electrocardiogram and electroencephalogram equipment to measure the narwhal's heart rate and brain activity in response to a stimulus presented to its tusk after the animal had been captured and held on a beach. When the narwhal shook his head from side to side and his heart rate escalated when highly saline water was pumped across his tusk, Nweeia was convinced that the tusk could sense variations in salinity levels and perhaps all sorts of other unknown stimuli.

×　×　×

Few biologists concur with Nweeia's conclusion that the narwhal's tusk is a sensory organ, and many are quite critical. They believe, instead, that the tusk is a secondary sexual characteristic, like the mane of a lion, the antlers of deer, elk, and other ungulates, or the tail feathers of a male peacock. All have evolved, they say, as a means of promoting one's genes by demonstrating their fitness to prospective mates.

The biologists point to English naturalist Charles Darwin, who in his *Descent of Man, and Selection in Relation to Sex*, argued that the narwhal tusk and all other species' male adornments were for maintaining social rank and competing for females. "When the males are provided with weapons which the females do not possess, there can hardly be a doubt that they . . . have been acquired through sexual selection," he wrote in 1871.

Half a century before Darwin, Arctic explorer William Scoresby wrote in *An Account of the Arctic Regions with a History and Description of the Northern Whale-Fishery*:

> The use of the tusk in narwals [sic] is ambiguous. It cannot be essential for procuring their food, or none of them would be without it: nor is it, perhaps, necessary for their defense, else the females and young would be subjected to the power of enemies without the means of resistance, while the male would be in possession of an admirable weapon for its protection. Dr Barclay, with whom I have communicated on this subject, is of opinion that the tusk is principally, if not solely, a sexual distinction, similar to what occurs among some other animals. Though it cannot be essential to the existence of the animal, it may, however, be occasionally employed.

In interviews with several biologists about Nweeia's work, all agree that his dental analysis of the tusk's evolution and structure contribute valuable information to understanding the narwhal's life history, but all dismiss as a leap of faith his claims about its function, calling it a "crazy theory," "delusional," or "yet another crazy hypothesis to justify research funds." University of Washington biologist Kristin Laidre said, "The narwhal's tusk is an almost exclusively male feature, a secondary sexual characteristic used for social interactions, establishing dominance, and competition for females between males. This occurs all over the animal kingdom." She said that she commonly observes male-male narwhal behavioral interactions using the tusk that are clearly social and in association with a nearby female. She suggests that the idea that the tusk is a "hypersensitive survival organ used to navigate the environment and find food" or anything other than a sexual trait "crumbles under the merest scientific scrutiny." Any trait critical to survival, especially with respect to navigating an environment, would most certainly also be found in the female portion of the population, which carries the added burden of reproducing and caring for young, she argued.

Perhaps the harshest critic of Nweeia's sensory organ theory is independent biologist James Finley. "The romance of the narwhal makes people want to fantasize all sorts of bizarre function. Nweeia is dreaming," he wrote to me during a month-long email exchange on the subject. "It continues to amaze me that practically every account of the narwhal, even some by biologists, has to end with the assertion that we still don't know the function of the mysterious narwhal tusk. It's no more mysterious than a moose's antlers, yet we are loath to let the legend die. So, hell, there's always room for another wild theory." He then shared with me a humorous poem attributed to Max Dunbar, whom Finley called the dean of Arctic marine biology, as an equally outrageous theory explaining the purpose of the tusk:

> Polar bears and narwhals gambol
> in the arctic sun,
> They frolic 'mid the ice floes in
> pursuit of arctic fun.
> The narwhal's tusk? Well, since you ask,
> the narwhal, when it dares,
> Employs its tusk, mischievously, in
> goosing polar bears.

Finley had heard about Nweeia's Arctic field research through colleagues who met him there, and later he saw the splash the dentist made in newspapers around the globe following a presentation at a marine mammal conference in San Diego in 2005. Finley said that the news accounts made him laugh and sigh and prompted him to write a letter to the editor of the *New York Times* debunking the dentist's proclamation. "The manner in which the press jumped on it and the way in which Nweeia exploited his moment of fame was predictable because that's what the newspapers want—mystery over fact—it sells papers, and is how you generate research funds," Finley wrote to me later. "I'm all for conducting any sort of esoteric research, but when you hype it and sell it to the public like this fellow did, I find it rather odious. I resign myself to the fact that the public doesn't want to see myths dispelled, and reporters and authors make their living by it."

Laidre agreed, noting that "the bigger problem is that respected media outlets reported this fringe hypothesis as a scientific discovery despite the fact there is no support for it whatsoever. In doing that, they have blurred the line between themselves and a supermarket tabloid. The misinformation has propagated in magazines, film documentaries, and even respected museum exhibits. Frankly, it is irresponsible," she added.

The news stories Finley and Laidre refer to include reports in many of the most prestigious newspapers in the world. Nweeia was quoted in the *New York Times* as saying "this whale is intent on understanding its environment" and theorizing that the tusk may help narwhals sense water temperatures, determine if ice is freezing, or track environments that favor their preferred foods. Yet in my interviews with him, Nweeia denied ever suggesting that the tusk was vital to the survival of narwhals.

At that, Finley responded with even greater passion, suggesting that Nweeia is "trying to cover his ass. If it is as sensitive as he says, then there's a hell of a lot of narwhals out there that must be screaming in pain from broken tusks (yet no one has documented a rogue narwhal)," he wrote. "It would have been best if he had just described the enervation [and] compared it to other species. The porosity of the tusk might be explained by the need for flexibility in such a long tooth used as a trysting pole."

Finley concluded his rebuttal of Nweeia's theory by returning to the text of William Scoresby, who, in writing about the bowhead whale, noted: "Large as the size of the whale certainly is, it has been much over-rated; for such is the avidity with which the human mind receives communications of the marvelous, and such the interest attached to those researches which describe any remote and extraordinary production of nature, that the judgment of the traveler receives a bias, which, in cases of doubt, induces him to fix upon that extreme point in his opinion which is calculated to afford the greatest surprise and interest."

Martin Nweeia is undeterred by the abundance of criticism he has received about his hypothesis from the community of biologists. He is absolutely convinced he is right, and he isn't worried about reversing the opinions of his naysayers.

"I think people should think whatever they want to think and do whatever they want to do," he told me. "The interesting thing about science is that it's kind of like life. You're faced with your perception of what you think is happening, and you're faced with truth. . . . You can go around the world and you can take your perceptions and you can say that's truth. But really, if you're a truth seeker, you understand that it could just be your perception." In the case of the narwhal's tusk, he said that no one before him had ever subjected a narwhal tusk to a battery of experiments to determine if it had sensory characteristics, so it would be premature for others to be so certain that it did not.

As he argued his position, it seemed as if he were fighting a tidal wave of opposition with only his principles on his side. And while I give him considerable credit for standing up for his beliefs and putting the narwhal's tusk through that battery of tests, I couldn't help but wonder, as I read and reread the transcript of our conversation, if his own perceptions were getting in the way of the truth. And yet he remained convinced and convincing.

While Nweeia told me that he doesn't really care what the biologists think, I got the sense that he did care and that he would like to be more accepted by other narwhal researchers. Yet, that desire wasn't going to make him give in to their predispositions. In fact, it may have made him even more determined to prove to them that he is right. And he seems happy to try to prove it to them on their terms.

"If I could talk to them, I would say, 'Look, we can both agree that because [the tusk is] a trait that's primarily expressed by males, that it has to do with sexual selection,'" said Nweeia, who has never spoken to his detractors. But then he added that perhaps one factor that would give some males a greater advantage when competing for females is to know where the female narwhals can be found. What if the females, he wondered, need to be in water of a certain salinity level to be amorous, and what if the tusk is sensitive to that salinity gradient? Or what if the females give off a pheromone when they are ready to mate, and what if the tusk can detect that chemical?

"Would that not be a reason to have a tusk?" he asked. "You would be hard pressed to find a biologist who would not want to buy that argument as a possibility."

× × ×

Sorting through the arguments for and against the sensitivity of the tusk had my head spinning. I had been equally perplexed about what to expect on that first trip to Pond Inlet and Koluktoo Bay. Between uncooperative weather, communication problems, and logistical mismanagement, the trip didn't go as planned. It might well have ended right after that one spectacular day, as I couldn't have imagined anything topping it. And I was mostly right. For the next two days, the wind picked up slightly, the water conditions made narwhal observations difficult, and while we took the boat out for short trips, we only had a brief glimpse of a small handful of distant narwhals. Once, we traveled to the end of Milne Inlet where a mining company was loading iron ore onto a barge. Returning to camp, a flock of thirty rare Sabine's gulls flashed their black-and-white wings as they passed us going the opposite direction, then reversed themselves and winged by us again minutes later, disappearing through a dazzling horizon-to-horizon rainbow that lasted for nearly an hour. But there wasn't a hint of a narwhal.

When the time finally came to depart for home, we were eager to leave our campsite amid the high Arctic islands of Canada for the four-hour boat ride to the airport. Renay and I and our three traveling companions wanted to get an early start, but breaking camp took longer than we expected. We climbed aboard our guide's modest boat at 10:30 a.m. for the ride to Pond Inlet, stopped an hour from camp to refuel from a rusty barrel stashed against a sheer cliff, and then had an unexpectedly rough ride out of the inlet.

As we rounded the corner past Ragged Island and into Eclipse Sound for the final two-hour push back to what qualifies in the region as civilization, a large group of narwhals seemed to appear out of nowhere, moving quickly toward us. Although their tusks weren't visible, their large size and gray bodies with pale speckling suggested that most were adults. At least thirty were blowing and surfacing simultaneously in the distance, and there was lots of splashing as they arched high out of the rough water to breathe. There may have been as many as one hundred narwhals or more in the group. They appeared to be in a great hurry, with a certain destination in mind, all sprinting in the same direction with no time for feeding or lollygagging. Namen quickly stopped the boat, dropped anchor, and offered us lunch while we enjoyed the encounter. As the whales approached moments later, most of them avoided us by staying underwater or detouring around us, so we didn't get great views of them. But as we chewed on our sandwiches and looked behind us, all of them resurfaced in a huge splashing herd and kept moving into the inlet.

We watched as a few more typical-sized pods of four to six narwhals passed us headed in the same direction, when another huge congregation could be seen coming our way, just like the first ones—splashing, surfacing, breathing loudly, and traveling fast and straight. This time, though, when they got near us, they didn't veer off or remain submerged. One small grouping surfaced three times right off the starboard—mostly dark ones with pale patches—and they were so close we could see their tusks beneath the surface as they rose to catch a breath. Another small pod included a rather small narwhal just behind the leaders, darker than all the others, probably no more than four or five feet long. Many others surfaced ahead of us, then remained submerged as they passed by. They were so close that we could see them as a pale blur underwater, looking almost like their closest relative, the similarly sized beluga, the only truly white whale. I wondered if they were rolling on their backs, as they are sometimes known to do, and flashing their light bellies at us. I repeatedly spun around to scan the water, and others onboard called out new sightings with enthusiasm. Even Namen and his brothers were genuinely excited to be in the midst of such a large gathering, so clearly it was an unusual event. It took about four minutes for the second herd of one hundred or more narwhals to pass by, and then as quickly as they had arrived, they disappeared. We packed up our lunch and continued on our way.

Namen's father was on the radio as we left, and he reported what likely had caused the mass of narwhals to hurry off—a rare pod of killer whales was in the area. Five minutes later, when another call came in, the boat made a hard turn to the north and picked up speed. Moments after that, we converged with the boat carrying the *Survivorman* crew trailing the orcas. First we saw three killer whales surface one after another, then two more thirty yards behind the first, then three more behind them. Compared to the narwhals, they were enormous, with their tall, black dorsal fins giving us advance warning when they began to surface, and their gray saddle patches and white markings beneath their tails and jaws clearly visible every time they took a breath. No wonder the narwhals were so frightened. I was, too. There may have been as many as twelve or fifteen killer whales surfacing all around us, but I couldn't keep them straight as they changed directions and split around our boats. Then, as so often happens with whales, they simply disappeared, the whole encounter probably lasting less than one minute. As one last orca leaped in the distance behind us, we turned for home. We were tired and sore and smelly and wind-burned, but that last show of narwhals and killer whales wrapped things up rather nicely.

×　×　×

The surprising sighting of killer whales chasing a large group of narwhals made me wonder how much of a threat the orcas are to the ice whale. It must not happen often, I guessed, since Namen and his brothers had never before seen a killer whale despite having lived their entire lives in the area. Like most questions about wildlife in the Arctic, answers aren't easy to come by. While killer whales are considered resident in the Canadian Archipelago when the water is ice free, their numbers are small and they are seldom observed because they are spread out over a wide area. Despite their small numbers, however, killer whales may be the top predator on narwhals—next to humans— according to many Arctic marine mammalogists, including one who referred to narwhals as "orca candy."

"My general feeling," said Kristin Laidre, "is that the densities of killer whales in the Arctic are low. Sightings of killer whales are, in general, pretty rare. We do know that they feed on narwhals and belugas, and some killer whale pods, it seems, have evolved to know precisely where narwhals are located in summer, especially in the southern part of their range like in Foxe Basin or Hudson Bay. There they show up pretty regularly, as narwhals are a predictable prey resource that occurs in high densities in ice-free shallow waters." Killer whales are sighted only rarely along the coast of West Greenland, she added, noting that when they are, it is talked about in the community for weeks.

Laidre is one of very few biologists to have observed orcas feasting on narwhals. In August 2005, while satellite tagging narwhals in Admiralty Inlet in northwestern Baffin Island, she and two colleagues watched as a pod of twelve to fifteen killer whales attacked and killed at least four narwhals among a group of several hundred over a six-hour period. From their observation point at Kakiak Point, they saw what they described as "vigorous surface and diving activity" by the orcas which resulted in a large oiled area on the water, presumably from whale oils released from the dead narwhals, and congregations of seabirds. It appeared that the orcas consumed the narwhals below the surface. The biologists had tagged several narwhals a few days before the attack, so they were able to monitor the movements of the animals in response to the killer whale aggression. According to Laidre, the narwhals in the area suddenly moved into shallow water as the killer whales approached, some forming tight groups and others lying still at the surface or moving slowly and quietly. One narwhal even stranded itself on a beach and thrashed its tail violently for thirty seconds as if to warn its pod mates. During the attack, most of the narwhals in the area moved as much as fifty miles south and spread out much more than usual. The animals resumed their normal behaviors within an hour after the killer whales departed the area.

In a surprising coincidence, given how seldom killer whale attacks on narwhals are observed, another biologist watched killer whales prey on narwhals on the exact same day in Repulse Bay, about 400 miles south of where Laidre made her observation. Laidre surmised that if the predation level from these two attacks were representative of the daily activity of killer whales in the region, then 200 to 300 narwhals are likely killed on their summering grounds by orcas during the two months of open water in the area. Coupled with the annual harvest by Inuit hunters and predicted reductions in sea ice, enabling killer whales to hunt narwhals over a longer period each year, this mortality rate raises questions about how these elements will affect the sustainability of narwhal populations.

×   ×   ×

Back home in Rhode Island, finally warm and dry, I couldn't get the tusk debate out of my mind. It got me thinking about the other debate that involved the narwhal's tusk, a debate that lasted for centuries but which is now considered a fantasy.

I have difficulty understanding today's fascination with unicorns (though I'm sure that plenty of others would say the same thing about my passion for collecting animal skulls). One blogger wrote that unicorns "embody the soul of every woman on earth" and are "mysterious and attractive, bold and unafraid yet shy and timid, contradictory and caring all at once." No they're not, I wanted to tell her. They're a figment of your imagination and have none of those qualities. But it's a perception that I'm not going to even attempt to fight. Tens of millions of pages of the Internet are devoted to unicorns, including more than three million artistic renderings, most in pastel colors, and thousands of figurines, stuffed animals, and other unicorn products are available for purchase. Who knows how many devoted fans have unicorns tattooed on their bodies? And yet it is likely that few of them have even given the slightest consideration to its origins and the role played by the narwhal in the mythology of the unicorn.

"The unicorn in your head depends on your background," wrote Christopher Lavers in 2009 in the introduction to his excellent treatise *A Natural History of Unicorns*. "If you are steeped in the Christian tradition you may conjure up a small, gentle, goat-like animal, perhaps cradled in a woman's lap. If your upbringing was secular you may imagine a muscular, noble, equine creature, perhaps facing a lion on a coat of arms. If you have read your children to sleep or dallied with New Age mysticism your unicorn is most likely a soft-focus, airbrushed, magical beast, perhaps with a crystal about its person. At once we see that the unicorn has different faces." And one of those faces is that of a narwhal.

The first written record of a unicorn appeared in about 398 B.C. from Ctesias of Cnidus, a Greek physician from what is now southwest Turkey. The vivid description of a unicorn that he provided in his book *Indica* noted that the animal was found in India and it resembled a wild ass, larger than a horse, having a white body, dark red head, dark blue eyes, and a horn on its forehead extending a foot and a half in length. It notes that the base of the horn "is pure white; the upper part is sharp and of a vivid crimson; and the remainder, or middle portion, is black." He went on to say that those who drink out of cups made from the horn are not subject "to convulsions or to the holy disease [epilepsy]. Indeed, they are immune even to poisons if, either before or after swallowing such, they drink wine, water, or anything else from these beakers."

While most scholars and others have long believed that Ctesias simply imagined the creature he described, or worse, was a liar, Lavers, a professor of ecology and biogeography at the University of Nottingham, took a different approach. He spent considerable time in his book attempting to trace the lineage of the animal that Ctesias wrote about back to the actual species that the man may have been describing. After first following the detective work of Odell Shepard, a Pulitzer-prize winner who concluded in 1930 that Ctesias's unicorn was a composite of three little-known but real animals, Lavers took the analysis a step further and concluded that Ctesias had merged elements of the Indian rhinoceros, the Tibetan antelope, and the kiang—a large, wild Indian ass—to arrive at his unicorn.

Lavers drew this conclusion, in part, from these facts: the single horn of the rhino was often made into drinking vessels and was presumed to have properties to neutralize poisons; in profile, the Tibetan antelope or chiru appears to have just one horn (though in fact it has two), and is believed to be the basis of a Himalayan legend about a one-horned beast, and its horns were also thought to have medicinal properties; and the kiang provides the horse-like elements of Ctesias's description, and its coat in summer is patterned red and white.

Whether or not these three animals were the basis of the mythical unicorn, later writers—including Aristotle, Pliny the Elder, and Claudius Aelianus (whose unicorn, Cartazonus, had a horn that grew in spirals or rings and tapered to a sharp point)—kept alive Ctesias's claims about the unicorn for several hundred more years. Aelianus' description of Cartazonus may be the first reference to the type of horn that became common on unicorns illustrated in art a thousand years later and became linked to the narwhal tusk.

Oddly enough, the unicorn also found its way into the Bible, though it took a far different path to get there. According to Lavers, the seven references to unicorns in the Old Testament likely appeared as a result of a series of mistranslations. Early Hebrew texts referred to a large, horned animal called a reem, which, when translated into Greek and later anglicized, became monoceros, meaning one-horn. "One-horn in Latin is unicornus, and unicornus in English is unicorn," wrote Lavers. "The reem of the ancient Hebrews became a monoceros in the Greek Bible, a unicornus in the Latin Bible, and eventually a unicorn, einhorn and so on when the bible was translated into the vulgar tongues of Europe. Ctesias's ass, or a memory of it, wound up in arguably the most influential book in human history."

So it is no wonder that for many centuries, unicorns were thought to be real. Even ancient naturalists (and some not so ancient) included the fabled animal in their catalogues of wild creatures, adding further credence to the story of the unicorn.

Religious scholars and art historians have long considered the unicorn to be a symbol of Christ, and artworks and tapestries up through the Middle Ages were filled with unicorns illustrating symbolic stories about Jesus. Perhaps the best known are the Unicorn Tapestries, woven between 1495 and 1505 and now on display at the Cluny Museum in Paris and the Cloisters Museum in New York. The unicorn's popularity as a religious symbol lasted for more than 500 years and reached its height in the fourteenth and fifteenth centuries, but it fell out of favor in the middle of the sixteenth century, due in part to unicorn lore expanding into the secular realm. It was the unicorn's spiral horn that came to be its defining characteristic.

×   ×   ×

Somewhere around the year 1000 A.D., the first narwhal tusks began making their way to Europe. It is believed that the Vikings once hunted narwhals, and they may also have acquired narwhal tusks from native hunters in the Arctic. George Frederick Kunz, in *Ivory and the Elephant*, wrote that the Vikings "decorated the prows of their war galleys with these horns, had them carved into sword and dagger hilts, and also set them on staffs and scepters. Their wives wore hair-pins made out of this material, and curiously wrought charms, which were considered talismans of good luck both in love and war."

The belief that unicorns were real and that the ivory shafts of narwhal tusks turning up in Europe, the Far East, and elsewhere were proof of their existence, supported the development of a high-end market for narwhal tusks. Unscrupulous traders, who knew that the tusks came from narwhals and were not unicorn horns, sold them as such and earned a hefty income. Odell Shepard estimated that just fifty whole tusks, each with a complete provenance, existed in Europe around the mid-sixteenth century, though many more were sold in pieces or powdered. As William Broad wrote in the *New York Times*, they became an icon of power, both for their religious associations and as a symbol of great purity, worth many times their weight in gold.

For instance, in the sixteenth century, Queen Elizabeth I received a tusk valued at £10,000, which was considered the cost of a castle at the time. A tusk owned by the King of France around the same time was valued at £20,000. English dramatist Thomas Dekker wrote in 1609 that a single narwhal tusk was worth "half a city," though what that truly equaled is uncertain. And according to Austrian lore, Kaiser Karl V paid off the nation's debt with two narwhal tusks. The Austrian royal family also had a narwhal tusk made into a scepter extensively decorated with diamonds, rubies, sapphires, and emeralds. The royal coronation throne of Denmark, at Rosenberg Castle in Copenhagen, was made of narwhal tusks, with Christian V becoming the first Danish king crowned in this chair in 1671. The throne is a highlight of castle tours even today.

The material that a narwhal tusk or supposed unicorn horn was made of was often called alicorn by those who sold or traded it in Europe, and while it was a sought-after commodity passed from royal family to royal family, its greatest value was perhaps its perceived usefulness in curing the ill or protecting one from evil. Eager to foil assassins, many kings and other leaders had goblets made of the precious material, which was believed to perspire when containing a tainted drink. In powdered form, it became much more available to commoners since fake alicorn powder could be made from bones, hoofs, and other abundant materials instead of real narwhal tusks. Churches put powdered alicorn in holy water to offer their ailing parishioners the hope of miracle cures. A London doctor advertised a drink made from powdered alicorn that could cure scurvy, gout, ulcers, consumption, fainting, rickets, and melancholy, among many other ailments. Others claimed it served as an aphrodisiac or as a means of determining whether a woman was a virgin.

But during the age of expanded Arctic exploration, the deception became evident. In 1621, cartographer Gerhard Mercator reported that narwhal tusks were the source of the purported unicorn horns. Seventeen years later, Danish zoologist Olaus Wörm investigated the matter further and exposed the ruse in a public lecture, but it was still one hundred years or more before the sale of alicorn went out of fashion. While Wörm made it clear that the unicorn horns and alicorn sold and treasured throughout Europe came not from unicorns but instead from narwhals—soon after called sea unicorns—this conclusion "bolstered the view that unicorns existed," wrote Chris Lavers, "because at the time it was widely believed that all terrestrial animals had marine counterparts. If unicorns existed in the sea, it stood to reason that they also existed on land. And so the unicorn refused to go away."

Did the Inuit or Viking hunters know that the narwhal tusks they traded to Europeans were later sold as unicorn horns? Probably not, according to Richard Ellis, whose 1991 book *Men and Whales* examined the narwhal-unicorn connection. He is convinced that the Inuit had never heard about unicorns, so they weren't hunting narwhals to provide tusks for the unicorn market, nor did they know that some people believed that narwhal tusks had magical or medicinal properties.

During the period of Enlightenment, when rational thinking dominated intellectual discussions, speculations about the mythical unicorn were ignored. But in the following Romantic era, the unicorn legend reappeared, and expeditions throughout Central Asia, the Himalaya region, and Africa were undertaken to seek evidence that the animal was real. The last such mission—this time to Uganda—was launched in 1899 by that nation's British governor, Harry Johnston. At its end, any remaining hope of finding a unicorn had disappeared, because it was believed that the planet had been too well explored to have missed finding a unicorn.

That's not to say, of course, that the narwhal tusk didn't remain a mystery. In many people's minds it still does. Unfortunately, the mythology of the unicorn has led to the somewhat common belief that the narwhal is a myth as well. To me, that raises a concern that this misconception may mean that the whale might not receive the protection it deserves from the very real and growing threats it faces.

FIGVRE DV POISSON NOMME VLETIF,
espece de Licorne de mer.

A sea unicorn in *Discours d'Ambroise
Paré* by Ambrose Paré (1582). The
animal's tusk or horn looks like that of a
sawfish, and early explorers thought the
two mammals were closely related. New
Bedford Whaling Museum Research
Library.

Male narwhals surface to breathe in Baffin Bay, where most narwhals spend the entire winter in small openings in the sea ice. Paul Nicklen/National Geographic Stock.

Perhaps the first illustration of a narwhal hunt, in *A New Voyage to the North* by Pierre Martin de la Martiniére (1700), an image described as being without precedent at that time. New Bedford Whaling Museum Research Library.

Between dives narwhals often remain still at the surface, replenishing their oxygen stores before diving again to search for food. Flip Nicklin/National Geographic Stock.

Polar bear skin on laundry line in Pond Inlet, Nunavut. Photograph by Renay McLeish.

In Koluktoo Bay, a tunnel-like iceberg almost as large as a city block. Photograph by Renay McLeish.

The royal coronation throne of Denmark in Rosenberg Castle, Copenhagen, is made largely out of narwhal tusks. Photograph by Renay McLeish.

The village of Qaanaaq, Greenland, is the northernmost municipality on Earth, where its six hundred residents largely subsist on local animals, including narwhals, seals, fish, walruses, and polar bears. Photograph by the author.

Small boats sit in the harbor in Qaanaaq. Photograph by the author.

The tail of the narwhal has a unique
heartlike shape. Paul Nicklen/National
Geographic Stock.

Hunter Mads Ole Kristiansen throws
his harpoon at a narwhal near the hunting camp of Siunnertalik, a two-hour
boat ride from Qaanaaq. Photograph by
the author.

After a successful hunt, Mads Ole Kristiansen poses with his prize, a nine-foot-
long female narwhal about four or five
years old. Photograph by the author.

While feeding in the winter, narwhals dive 1,800 meters to the seafloor, a round-trip that takes them about thirty minutes. Paul Nicklen/National Geographic Stock.

Icebergs calved from seven nearby glaciers float out of the Inglefield Bredning fjord and pass by Qaanaaq. Photograph by the author.

The research camp on the shore of Tremblay Sound on Baffin Island, Canada, where scientists lived for three weeks while trying to capture narwhals. Photograph by the author.

After trapping a narwhal in a net, the researchers collect measurements and tag the animal so they can track its movements. Photograph by Cortney Watt.

The satellite tag attached to the whale's dorsal ridge collects data about the whale's location and diving depth. Photograph by Cortney Watt.

The research team at Tremblay Sound
included scientists, a student, camp
managers, and local Inuit guides. Photograph by Cortney Watt.

~~~ SIX ~~~

MELTING ICE

THE LINE OF EVOLUTIONARY DEVELOPMENT THAT LED TO
today's whales—and to humans as well—can be traced back to the
Cretaceous period and the carnivorous land mammals from which
they descended. The branch that became whales follows that of the
artiodactyls, hoofed mammals like antelopes and camels, as well as
the hippopotamus, the whale's nearest living relative on land. The discovery in 1978 of a 52-million-year-old skull of what became known as
Pakicetus was found to have features that showed a transition between
terrestrial mammals and aquatic animals, including modifications
that allowed for directional hearing under water, one of the first hints
that mammals were returning to the sea. But there were numerous
additional steps before the first true whales emerged.

An amphibious animal called *Ambulocetus*, with hind feet clearly
adapted for swimming, followed *Pakicetus*, and later still came *Rhodocetus*, which had additional adaptations for a marine lifestyle, including legs disengaged from its pelvis. By the Eocene, about 40 million
years ago, *Basilosaurus* emerged as a fully adapted marine mammal,
with a streamlined body, paddle-like flippers, a strong whale-like tail,
and the remnants of hind limbs that soon would disappear. Other
physiological changes took place along the way as well, including the
relocation of the nostrils from the snout to the top of the head, the
addition of an insulating layer of blubber, and changes in the circula-
tory system for the management of oxygen and for withstanding the
pressures encountered at great depths.

The two groups of modern cetaceans, the toothed whales and baleen
whales, descended from a third group, the Archaeocetes, which disappeared about 30 million years ago. It was just 500,000 years ago that
narwhals evolved as a species, sometime in the late Pleistocene, about

the same time that polar bears diverged from brown bears and when many large mammals and birds evolved and went extinct. It was also a time when great changes in climate occurred. While always believed to be an Arctic species, the narwhal's range expanded and contracted over the millennia as variations in climate dictated. During the last glaciations about 50,000 years ago, when ice extended as far south as England, narwhals were forced southward. Fossils of the whales have been found there, along the coast of Norfolk. When the glaciers began their retreat, narwhals followed them north again, with some traveling to the east of Greenland and others to the west. That was the last time that narwhals from West Greenland and Canada came into contact with narwhals from East Greenland. The 10,000-year separation of the two populations has led to genetic differences between them, a signal that evolutionary changes are still taking place, changes that one day far in the future could result in two distinct species of narwhals. If they last that long.

What is most evident in their evolution, however, is the narwhal's direct link to the ice. Regardless of where the edges of the glaciers and the ice pack have been located—Greenland or England or somewhere in between—that's where narwhals seem to thrive. Even today, their southbound migration is triggered by the formation of ice as winter approaches, and they chase the retreating ice northward again in spring. Their preference for spending the winter months where no other mammal can survive, amid the tiny openings in the pack ice in the middle of Baffin Bay, is further evidence of their dependence on that icy realm, a place where humans would die in just a few minutes despite having a similar mammalian physiology.

With a layer of blubber three or four inches thick, narwhals are uniquely adapted to surviving in the extreme cold. At great depths, where it is often even colder but where they must dive to feed, they are faced with pressures greater than 2,200 pounds per square inch, which they withstand with a flexible and compressible rib cage that can be squeezed without harming them as pressure increases. To carry along enough oxygen to sustain them on deep dives of up to twenty-five minutes, they have evolved several nifty solutions, including an enormous concentration of oxygen-binding myoglobin in their muscles, more than twice that of most seals and eight times as much as land animals, enabling them to swim under water at speeds of one meter per second for twenty minutes without taking another breath. With muscles better suited for endurance than for speed, narwhals can save even more oxygen by turning off the blood flow to noncritical organs and other body parts.

These adaptations to living in a world of ice raise the vital question of whether the narwhal can continue to thrive when the ice disappears and the water warms. It's a question for which there is no answer, but it's a scenario that is quickly becoming reality. And it's a reality to which narwhals may be ill-equipped to adapt.

×　×　×

Polar bears have received most of the attention in the news in the last few years for the potential effect that climate change will have on their survival. As the planet warms and the sea ice shrinks and retreats farther from land, the ice platforms from which the white bears hunt for seals, which make up the bulk of their diet, become less accessible. The results have already come to pass, with bears swimming for days far out to sea, hungry and emaciated, trying to reach land or find one last bit of ice upon which to rest and seek a meal. Drowned bears, both adults and cubs, continue to wash ashore, and there are certain to be many more that sink or are consumed by other creatures and never counted. In addition, many polar bears are becoming landlocked, stuck on barren ground after the snow has melted and the icepack has withdrawn, with no way to reach the seals that they must bulk up on during the spring months to carry them through their dark hibernation. Some have been documented, for the first time ever, climbing rock cliffs to find and eat seabird chicks and eggs on Coats Island at the northern end of Hudson Bay.

Hybridization between polar bears and grizzly bears is becoming another climate-driven concern. DNA tests of a white bear with brown patches shot by hunters in 2006 confirmed that it was a hybrid. Another was killed in 2010, but this time it was a second-generation cross whose mother was a hybrid and father a grizzly. More hybrids were found in 2011 and 2012. While biologists knew the two species could mate and produce viable offspring—a hybrid bear in a German zoo exhibited seal-hunting tendencies but poor swimming abilities—it wasn't expected in the wild.

What does this have to do with climate change and the narwhal? A great deal. "Rapidly melting Arctic sea ice imperils species through interbreeding as well as through habitat loss," wrote marine biologist Brendan Kelly in 2010. "As more isolated populations and species come into contact, they will mate, hybrids will form and rare species are likely to go extinct. As the genomes of species become mixed, adaptive gene combinations will be lost."

Kelly has identified thirty-four hybridizations of Arctic and near-Arctic marine mammals involving twenty-two different species, two-thirds of which are listed on or candidates for the U.S. endangered species list. These include a narwhal-beluga hybrid from West Greenland documented in the 1980s. With the decrease in sea ice, polar bears will likely spend more and more time on land in the same areas as grizzlies. While scientists are uncertain how this interbreeding will affect populations, they know that when humans have introduced species into foreign lands in the past, hybridization has quickly reduced genomic and species diversity. Kelly is worried that at-risk species like polar bears will follow suit and breed themselves out of existence.

As worrisome as the warming planet is becoming for polar bears, a 2008 analysis of the affect of climate change on Arctic marine mammals concludes that while all seven species—narwhal, beluga, bowhead whale, bearded seal, ringed seal, walrus, and polar bear—are threatened to some degree by climate change, narwhals may face the greatest risk. Yet all species have experienced warm spells in their distant past and survived to spawn future generations.

To understand how Arctic climate has changed through history and why the Arctic is the first region of Earth to experience such changes—and ultimately to figure out what this all means for the future health of the narwhal—I visited with several scientists who I hoped would help me make the link between climate, ice, and narwhals. Since most of the threats that are predicted to face narwhals in coming decades are related to the warming of their environment—including changes in the food web, increased shipping and oil exploration, exposure to new diseases and toxins, and increased competition for food with commercial fishing operations—I felt that I needed a primer on Arctic climate changes before I took my next journey north.

×   ×   ×

My first inquiry took me a short distance across campus from my office at the University of Rhode Island's Graduate School of Oceanography to meet with Kate Moran, who may know more about the history of Arctic climate change than anyone. A marine geologist and ocean engineer who later left URI to serve as a science advisor to U.S. President Barack Obama on issues related to oceans and the Arctic, Moran led an international team of scientists on a technologically challenging effort—the first ever attempted—to collect sediment core samples from deep within the seabed along the Lomonosov Ridge about 150 miles from the North Pole. Their aim was to see what the accumulated sediments could tell them about Arctic climate patterns from millions of years ago in hopes of learning something about the dramatic climate changes now taking place on Earth.

Moran said it took her and Stockholm University colleague Jan Backman nearly a decade to plan the 2004 expedition, which required the use of three icebreakers, including a nuclear-powered ship from Russia. Two of these vessels provided a protective shield around the third, which served as the coring vessel. The biggest challenge was keeping the coring ship in a fixed location for days at a time amid heavy, moving sea ice as it drilled into the seafloor 1,200 meters below the sea-surface. The other vessels were used to protect the drillship, continuously circling around, to break up and push away the three- to ten-foot thick sea ice.

The $12.5 million, four-week expedition involved more than 200 people, including ice management experts, engineers guiding the drilling process, scientists, ship's crew, and many more. The bulk of the scientists were those Moran called "the Noah's ark of micro-paleontologists," experts in various categories of ancient microorganisms, such as diatoms, forams, dinoflagellates, ostracods, and radiolaria, who could trace the comings and goings of these tiny creatures. "We had two of every kind of micro-paleontologist available because nobody had ever been to the Arctic before and we didn't know what kind of microorganisms we would find," explained Moran. Since there are known time periods during which various kinds of these organisms lived, the appearance of their fossilized remains in the sediment core was an indication of the period of history that the scientists had reached in the drilling.

The first major discovery of the project occurred during the drilling itself, when the drill rig struck bedrock beneath the sediment layer and it was determined that the bedrock was part of the continental crust that had broken away from the Barents Shelf over 60 million years ago. This conclusion, that the Lomonosov Ridge in the middle of the Arctic Ocean was once a part of the Asian continent, provided justification to the government of Russia to claim the Arctic—and all of the undiscovered oil and other natural resources it may contain—for itself. Soon after, Russia sent a submarine to the North Pole to sink a Russian flag on the seafloor and declare its sovereignty over the region. While the political posturing over who controls the Arctic will likely last for decades, the geologic discovery made by Moran and her team of scientists may carry considerable weight in future discussions.

As for shedding light on ancient climate, the expedition was nothing short of a stunning success. News media from around the world reported on every discovery, and it was the lead story in the most prestigious research journals again and again. Prior to this project, all that was known from sediment cores about the climate record in the Arctic came from a couple of ten-meter cores that revealed insights as far back as the relatively recent mid-Pleistocene epoch, about 200,000 to 500,000 years ago. The 400-meter sediment core recovered by Moran's team revealed an amazing 56-million year record of climate changes that could help clarify present and future climate trends. It exposed a surprising Cenozoic Era record of a climate transition from a warm "greenhouse" world in the late Paleocene and early Eocene epochs to a colder "icehouse" world influenced by sea ice and icebergs from the middle Eocene to the present.

For instance, the researchers discovered that 55 million years ago, during a period called the Paleocene Eocene Thermal Maximum, which some suggest is analogous to today's world because of a rapid release of greenhouse gases at the time, the surface temperature of the Arctic Ocean was much warmer than previously believed—perhaps as much as 20°C higher than today. Moran called this time frame the Big Heat, because it is the largest known climate warming of the Cenozoic, and because it was dramatically warmer than any climate models had predicted.

They also found the remains of vast quantities of a freshwater fern called *Azolla*—an ancient cousin of the duckweed found today in water gardens and suburban ponds—in sediments from 49.5 million years ago, which suggests that the ocean had considerably lower salinity levels back then. "The plant was still preserved as organic matter in the sediment, so that means it was probably really prevalent all over the place," Moran said. "It means that the Arctic Ocean was probably the biggest freshwater lake in the world." One thing the *Azolla* discovery means is that the plants growing on the sea surface would have absorbed carbon dioxide from the atmosphere and could be a response to warm periods. It also means that the region's hydrologic cycle was very different. Today the Arctic is considered somewhat like a desert, but it was probably quite lush during this ancient period. About 800,000 years later, when shifting land formations reconnected the Arctic with the Atlantic, warm salty water began to flow into the Arctic Ocean again and killed off the *Azolla*.

The sediment core recovered by the scientists also suggests that the first evidence of ice in the Arctic Ocean was 45 million years ago, which is about 35 million years earlier than previously believed. The evidence of the cooling climate comes from "ice-rafted sediment," pebbles and sand that fell out of floating ice. Because some of those pebbles came from hundreds of miles away from where they were found, from locations where it would have taken several years to transport, the sea ice probably existed year-round rather than just during the winter season.

"Previous to our expedition, the previous cooling period was thought to have begun about seven million years ago," explained Moran. "People thought the northern hemisphere cooled much later than Antarctica, which cooled 45 million years ago. But we found evidence of cooling in the Arctic 45 million years ago, which suggests that the poles may have been cooling at about the same time. And that could mean that the mechanism for cooling is atmospheric," probably from changing concentrations of carbon dioxide and other greenhouse gases.

Most recently, the scientists have been studying what the core reveals about the last 18 million years of Arctic climate. According to Moran, it all contains ice-rafted sediment, and much of that sediment had to travel in the ice for at least a year or two to get from where it was collected along the coast line to where it was deposited on the Lomonosov Ridge.

"If it takes multiple years to take a rock from one place to another, then it means that there's perennial ice from year to year," she said.

"We found that there has been perennial sea ice for the past 13 million years at least. That's the permanent ice cap that we see on satellite images, which contributes to the Earth's albedo [reflectivity] and keeps the Earth cool. That ice is a piece of the climate system. That summer sea ice may be gone in ten to fifteen years. It's a major piece of the climate system that has been around for a long, long time keeping the planet cool, and it is at risk of being lost."

x   x   x

I found myself next in Denver, Colorado, where, adjacent to the Federal Building and attached to an intriguing U.S. Geological Survey facility containing a stunning array of highly detailed maps, a warehouse-like structure contains one of the world's most unusual freezers. It's not that the freezer itself is particularly odd, but what it contains certainly is. This nondescript building is the National Ice Core Laboratory, and the samples it contains provide another way of studying the history of Arctic climate. The repository houses an archive of the ice core samples that scientists have collected from deep bore holes in glaciers in the Arctic and Antarctic regions since 1958, preserving the integrity of the ice for future studies of climate and environmental conditions. More than sixteen kilometers of four- to six-inch diameter cores from more than 80 different locations are stored in 14,900 cardboard tubes, much like mailing tubes wrapped with an outer layer of aluminum for added protection.

Renay and I arrived on a late October morning in the midst of an unexpected heat wave. We were there to talk with Eric Cravens, assistant curator of the lab, but we had no expectation that he would be taking us into the freezer, let alone stay there for an extended period, so we had dressed for the hike into the hills we had planned for later that afternoon—jeans and cotton shirts. But soon after we arrived, Cravens put on a heavy fleece jacket, a thick plaid scarf and woolen gloves, apologizing as he did so that he had no warm clothes to loan us. And in we went.

The first room, which Cravens called the exam room, was permanently set at −25°F, and it had a door like the one that led to the walk-in freezer in the Kentucky Fried Chicken where I worked as a teenager. We stepped through and pushed past a veil of vertical plastic strips and were immediately struck by the force of the cold. I have experienced the opposite trauma walking from a comfortably air-conditioned room outside into hundred-degree temperatures, when the air tasted thick and the humidity weighed me down like I was carrying a stove on my back. It never occurred to me that the cold would have a similar effect. The coldest temperature I had previously experienced was about −5°F during a frigid winter when I was in junior high school, and then I only went into the elements wearing my warmest clothes. This freezer was breathtaking, and not in a good way.

That first moment in the exam room brought to mind Bill Streever's wonderful book *Cold: Adventures in the World's Frozen Places*, which I was in the middle of reading at the time, no doubt preparing myself for that exact moment. In its introduction, he wrote that "cold freezes the nostrils and assaults the lungs. Its presence shapes landscapes. It sculpts forests and herds animals along migration routes or forces them to dig in for the winter or evolve fur and heat-conserving networks of veins. It changes soils. It preserves food. It carries with it a history of polar exploration, but also a history of farming and fishing and the invention of the bicycle and the creation of Mary Shelley's *Frankenstein*. It preserves the faithful in vats of liquid nitrogen from which they hope one day to be resurrected."

That last image gave me a chuckle, because that's how Renay and I were feeling soon after we walked into the freezer, and we knew it was only going to get worse. I had been taking notes during our earlier discussions with Cravens, but thirty seconds after we crossed the threshold and experienced a hundred-degree drop in temperature, the cheap pen my plumber had given me froze solid. I didn't even dare trying to operate my digital recorder for fear all my files would be lost.

The room wasn't much larger than my living room, with metal tables set up in the middle as work space, a computer enclosed in a plastic protective case in the corner, and a few flags hung from the ceiling representing the countries from which the facility's ice cores originated. Cravens quickly ushered us into a phone booth-sized nook against one wall draped in a black curtain in which he had displayed a one-meter ice core lying on a light tray for better viewing. The core came from what he called GISP2D, a site from what is known as the Greenland Summit located near the center of the glacier-covered island. There were 3,037 similar cores from GISP2D, all a meter in length and just over thirteen centimeters in diameter, stored in the next room at the Ice Core Laboratory. All had been drilled sometime between 1989 and 1993. The core Cravens had pulled out for us to examine originated 1,841 meters below the surface of the glacier, and it represented ice that had built up on Greenland over a forty-year period about 16,400 years ago. Due to the compression of ice over time, a one-meter core from the same hole but 600 feet deeper would have covered a 200-year time span.

With the aid of the light tray, it was easy to note numerous clear and cloudy layers extending around the core, like Life Savers in a roll. Each pair of layers equated to a year of accumulating ice, with the clear layers representing the colder winter months and the cloudy layers containing dust particles blown around in the warmer summer months. Cravens said that the best ice cores are those where the layering is distinct, and GISP2D has perhaps the most distinct layering of all of the cores collected so far. "If the layers are in place, it means there hasn't been too much of a disturbance there, so you can go back in time pretty good and realize that you haven't lost the sequential stratigraphic accumulation." The annual layers tell the story of the Earth's climate, most notably the composition of gases in the atmosphere, but also captured in the ice is a record of events like volcanic eruptions, cosmic particles, and nuclear dispersals. An analysis of the atmospheric gases trapped in the ice not only reveals details about the climate record as far as 800,000 years ago, but it also provides insights into weather patterns, sources of moisture, the biological productivity of the ocean surface, and the frequency of catastrophic events.

Scientists from around the world come to this exam room to conduct a wide range of analyses of the ice cores: electrical conductivity tests to detect the presence of acids and aid in the determination of the exact age of the layers; stable isotope analyses to determine the temperature of the water when it originally evaporated from the ocean surface; direct measurements of the concentration of gases trapped in the ice; and measurements of the physical properties of the ice crystals and bubbles in the layers to discern a complete history of the glacial ice. As we left the room, Buford Price, a researcher from the University of California at Berkeley, was preparing to enter it with a high-tech laser device that he was going to use to detect biological molecules within some Antarctic ice cores.

A door in the back of the exam room led us directly into another freezer room where all 14,900 ice cores were stored. As we entered, we noted another drop in temperature, down to −35°F, a shock less forceful than the first but one that still sent a literal shiver throughout our bodies. Fifteen-foot-tall racks of core tubes lined the space from floor to ceiling in a number of rows, reminding me of the end of *Raiders of the Lost Ark*, when the government stored the Ark of the Covenant in a massive warehouse of identical-looking crates. Our view of the room was entirely in black and white—white walls, white metal racks, white overhead light fixtures, and yards and yards of silvery black tubes labeled with codes for the site, depth, and year the core was collected.

After only six minutes, Renay couldn't stop shivering, and Cravens could see that I, too, was well out of my comfort zone, so he escorted us quickly out of the room. As soon as we left, my glasses crusted over with frost—a warm air blower like those used to dry your hands in a public restroom was located just outside to rectify that situation—but the rest of me felt much better almost immediately.

Cravens said that ice cores are one of the only places where the constituents from the atmosphere from centuries ago can be directly measured. At the surface of a glacier, he told me, are open cell structures called "firn" through which air can circulate. As snow falls and gets packed down and compressed, the open cells become closed off. That's where the snow transitions from firn to ice. Depending on the location around the world, the depth at which that occurs can be anywhere between sixty and 120 meters beneath the surface. When the cells become closed, the atmospheric gases that existed around the time the snow fell become locked in the ice. By measuring the concentrations of gases and other constituents in the ice, scientists can reveal details of ancient climate. One of those details was that the average temperature in the middle of Greenland had only varied by about 2°C over the last 10,000 years, but there was a drop of about 5°C about 10,500 years ago and considerable variation in temperatures for the 100,000 years before that. What caused those dramatic changes is unknown.

I asked Cravens the difference between what can be learned from the ice cores stored in his laboratory and the sediment cores collected from the sea floor by Kate Moran and her team. "The difference is magnitude and time," he said. "Ice cores, on the outside maximum at the moment, can go back 800,000 years. What you get out of ice that you don't typically get out of marine bottom sediments is resolution. [Sediment cores] will go back to 55 million years, but they won't give you annual resolution."

× × ×

The significance of the ancient climate record that Moran's team collected and the more detailed recent record that the ice cores reveal comes from the influence that the Arctic Ocean has on global climate. To learn how that works (and, ultimately, its affect on narwhals), I left Denver to make one last stop, traveling forty miles northwest to Boulder, Colorado, to the National Snow and Ice Data Center at the University of Colorado. In a bland brick building a few miles from the school's main campus, I met Mark Serreze, the director of the center. Bespectacled and ponytailed, he leaned back in his chair and gave me a brief lesson in climatology.

"What we see today is the Arctic in the midst of a rapid change to something entirely new or something we haven't seen for many thousands of years, perhaps even hundreds of thousands of years," he said. "The Arctic system, like the rest of the planet, of course, is warming up. It has long been known that if we were to start to warm up the global climate through any mechanism—turn the sun a little brighter, put more greenhouse gases into the atmosphere, or whatever have you— we're going to see the effects of this warming first in the Arctic, and that's really where it's going to be most pronounced."

He went on to explain that these pronounced effects are a result of what he called Arctic feedbacks and climate forcings.

"Climate doesn't just change all by itself. It's not like Harry Potter flicks his magic wand and climate suddenly changes. Something has to force it," Serreze said. And while there is certainly considerable natural variability in weather from year to year, the key to understanding long term changes in climate is to identify what is forcing that change. Some ancient climate changes have been traced to periodic changes in Earth's orbit, for instance. The Little Ice Age in the seventeenth century is associated with a period of reduced solar output and an extended period of volcanic activity that spewed particles into the atmosphere and cooled the planet.

"Now we've got a new forcing at work," Serreze said, referring to the emissions of greenhouse gases into the atmosphere, primarily from the burning of fossil fuels, which he said changes the radiation balance of the planet, forcing the system to warm up. How much it warms up depends on a variety of Arctic feedbacks, especially what he called "the albedo feedback."

Albedo is simply a fancy word for the reflectivity of a surface. White snow has a high albedo—80 percent of the sunlight that strikes snow is reflected back up into space. If the planet begins to warm up and some of that snow and ice begins to melt, it exposes darker underlying surfaces of ocean water or tundra, which have a lower albedo and absorb more solar radiation. That warms the system even more and melts more snow and ice, creating a domino effect of increased melting that cannot be stopped. Because the albedo feedback is mostly taking place in the polar regions, it's the Arctic and Antarctic that are being affected most by global warming. As warming occurs over the Arctic Ocean, that warmth spreads over land, hastening the decay of the permafrost in the tundra, triggering tiny organisms in the soil to become active and begin releasing some of the 1,600 gigatons of carbon stored there (compared to the 730 gigatons in the atmosphere), further aggravating the situation.

Serreze said that a number of other events have taken place in the last thirty years that have helped to accelerate the warming trend, including a change in atmospheric circulation from the late 1980s to mid-1990s that flushed out much of the thick ice from the Arctic into the North Atlantic, where it melted and left the Arctic with thinner ice. Changes in ocean circulation may also play a role, he noted, though the primary force is the albedo feedback responding to warming from a build-up of greenhouse gases.

The key question for narwhals, because their winter feeding grounds are 98 percent ice covered, is how long will sea ice remain in the Arctic. Serreze's estimate for that date has changed dramatically in just the last few years.

"If you had asked me that five years ago, the number I would have said is 2070 or maybe even around the year 2100, because that's what the climate models were telling us," he said. "But what we've seen since then is that the climate models are too slow. . . . Lots of people are working on this, but our view on it now is that we'll lose the summer ice cover by 2030 or so. Somewhere around there seems reasonable."

He added that the term "ice free" is an inexact statement, because it's likely that there will always be some ice in the Arctic, but he predicts that you'll be able to sail across the North Pole less than twenty years from now. It will look like a blue ocean at that time. Other reputable scientists are predicting the region will be ice free in half that time. To Serreze, the question of when it's going to happen is less important than the bigger issue of how that loss of sea ice will affect the Arctic ecosystem and the planet as a whole.

Scientists agree that the loss of sea ice will most definitely cascade down the food chain and may affect almost every species of marine life because the ice is the foundation of the food web. In the Arctic spring, hundreds of species of microscopic predators collectively called zooplankton begin to feed on tiny marine plants known as phytoplankton, which grow, fueled by the sun, on the underside of the ice. The predators, mostly amphipods and copepods—tiny crustaceans smaller than a grain of rice—are fed upon by bowhead whales, Arctic cod, and other fish, and the fish are eaten by seals, beluga whales, and narwhals. Polar bears, hunting from atop the receding sea ice, feed on the seals and, occasionally, on the whales. If the sea ice disappears, will the phytoplankton blooms still occur? And if so, where will they take place and how long will it take the populations of fish and whales to find them?

The good news for narwhals is that they eat very little during the summer, and they are only associated with ice in the winter, when Serreze said sea ice is not in jeopardy. While ice extent is declining in all months, he said it is occurring at a slower rate in the winter. September, at the end of the melting season, has had a decline in sea ice extent of 11 percent per decade since 1979, when satellite data begins, while the winter sea ice decline has been at a rate of just 3 or 4 percent.

"Even in the greenhouse warm world, it stays cold enough in the winter for ice to form, but the ice that forms is thinner than it was before," Serreze said. He doesn't expect that the maximum extent of sea ice in winter will change significantly in the next fifty to 100 years.

In fact, scientists have noted that in Baffin Bay and the Davis Strait, where the majority of narwhals spend the winter, the extent of winter sea ice actually increased until the late 1990s, perhaps the only place in the Arctic where that occurred. Since 2000, however, sea ice in Baffin Bay has followed the declining trend that other Arctic regions have been experiencing.

× × ×

It isn't just the disappearance of sea ice caused by climate change that will negatively impact narwhals and other Arctic marine life. Kristin Laidre, the leading narwhal biologist in the United States, said that the effect of climate change on narwhals could come from a wide range of factors that will result from a warming planet. The increased variability in the timing of sea ice formations or in the availability of leads (long, narrow openings in sea ice used for breathing in winter), for instance, will likely have negative consequences for the whales. Climate models predict that precipitation will increase, leading to more runoff from land, potentially affecting coastal water quality. Changes in ocean temperatures, salinity, or currents might alter the distribution of prey. Human factors, such as increased oil exploration and drilling, or increased shipping through the Northwest Passage, could disrupt migratory routes and feeding areas. "If you're a species that relies on specific predictable prey resources and you go exactly where they are found, if something happens and the system changes, you have to be able to adapt your behavior," said Laidre. "To some extent, these indirect impacts may make narwhals more vulnerable than the direct impacts of sea ice loss."

Adaptability to changing conditions seems to be a vital factor, and narwhals don't appear to have that characteristic programmed in their genes. Laidre led a team of biologists who conducted an analysis to identify the factors that would make an Arctic marine mammal vulnerable to climate-induced changes in its environment, and she found the narwhal to be among the top species facing the greatest threats. Their "sensitivity index" included such elements as worldwide population abundance, size of geographic range, reproductive potential, diet diversity, site fidelity, migratory behavior, habitat specificity, and sensitivity to changes in sea ice. "If you're flexible and adaptable and can deal with a wide range of habitats and situations, then you're a lot less vulnerable," Laidre said. Unfortunately, the narwhal scored poorly in all of those categories.

According to Laidre, narwhals are the most specialized Arctic cetacean and the most restricted in their distribution. What this means is that narwhals have very specific habitat and dietary needs and they are one of the least adaptable Arctic marine mammals, with the possible exception of the polar bear. It also means that even modest changes to their environment could have serious effects on their health and abundance. And climate change is almost certainly going to cause more than modest changes. For instance, a warming climate will cause many Arctic marine species to shift their populations northward, just as the cooling period of the Little Ice Age expanded the narwhal range southward as far as England in the 1700s. Northward expanding ranges of more southern species may mean that subarctic species could end up competing with narwhals for food and potentially altering predator-prey relationships.

In addition, researchers at the Scripps Institution of Oceanography have noted that the disappearing sea ice in the Arctic has triggered plankton blooms, which provide the basis of the food web in the region, to occur up to fifty days earlier than they did in the late 1990s. In Baffin Bay and Foxe Basin, for instance, blooms that used to peak in September have now shifted to early July. The timing of the availability of food is directly linked to the reproductive cycles of many animals, including marine mammals.

Changes to or loss of habitat may also place additional physiological demands on whales, which could lead to increased vulnerability to disease. Kathy Burek of the Alaska Veterinary Pathology Services led an examination of the effects of climate change on the health of Arctic marine mammals, and she noted numerous ways that narwhals and other wildlife could have their health compromised as a result of the changing climate. In 2008 she reported that climate warming may alter the relationship between marine mammals and infectious disease pathogens, including changes in pathogen transmission rates due to changes in movement patterns or marine mammal density; changes in susceptibility to disease due to nutritional or toxic stresses; and range expansion and increased survival rates of pathogens that would ordinarily die off in extreme weather. She noted that the current warming trend is linked to a worldwide increase in reports of diseases affecting marine organisms, including five separate incidents of mass die-offs of seals. Burek also worries that marine mammals will be increasingly exposed to a wide variety of toxins as a result of climate change, including increasing numbers of harmful algal blooms and the long-range transport of toxins in the atmosphere or in the ocean. These don't even include the anthropogenic factors caused by an increased human presence in the Arctic, like boat strikes, acoustic injuries, fishing gear entanglements, and noise pollution.

Changing ice patterns may also threaten narwhals in another way. According to a team of researchers at the University of California at Santa Cruz, when giant ice floes break away from glaciers—like the 250-square kilometer Petermann Ice Island that broke from a Greenland glacier in 2010 and drifted into narwhal habitat—they pose a hazard to the whales. That's because the scientists calculated that narwhals must surface for air every mile or so, meaning that they could easily drown if they wander too far beneath one of these massive icebergs. The researchers, Terrie Williams and Shawn Noren, reported that narwhals have muscles built for swimming long distances rather than swimming quickly, but the increasing abundance of large, highly mobile ice floes could be too large for them to swim beneath. "A wrong decision or a shifting wind moving ice could be fatal," they wrote.

So while narwhal populations are presently somewhat healthy, the warming climate makes their future precarious.

~~~ SEVEN ~~~

GREENLAND

IF I THOUGHT IT WAS AN ARDUOUS JOURNEY TO BAFFIN ISLAND, I was clearly unprepared for the challenges of getting to Qaanaaq, Greenland. The northernmost municipality on the planet, Qaanaaq lies on the northwest tip of the world's largest island and is home to about six hundred residents, the overwhelming majority of whom are native Greenlanders—Inuit, like their cousins across Davis Strait in Canada—with a smattering of Danes. To get there from Boston required seven flights over three-and-a-half days, including an overnight stay in the city of Ilulissat (formerly Jacobshavn), about half way up the island's west coast on Disko Bay. It is one of the only places where narwhals and belugas appear together during mating season, and it is where the only known narwhal-beluga hybrid was found.

Ilulissat sits amid rolling rocky hills and tundra with a harbor filled to the brim with small boats and a few commercial-sized fishing vessels. Travel literature and airport posters boast that Ilulissat glacier is "the most productive glacier in the northern hemisphere," by which they appear to mean that it calves a greater volume of ice into the sea than any other glacier. They say it like it's a point of pride or something to celebrate, but to me it's a worrisome feature and a further demonstration of the effects of global warming.

On the other hand, watching the icebergs float by my hotel room window was mesmerizing. I couldn't help but pull up a chair and stare out the window at the dazzling parade of ice late into the night. The largest iceberg appeared to be grounded near the entrance to the harbor, as that was the only one that didn't move, yet it was also the most impressive of them all—a somewhat square block with bluish horizontal striations on the side and a similar smaller block on top like the beginnings of a massive wedding cake. A few more squarish ice-

97

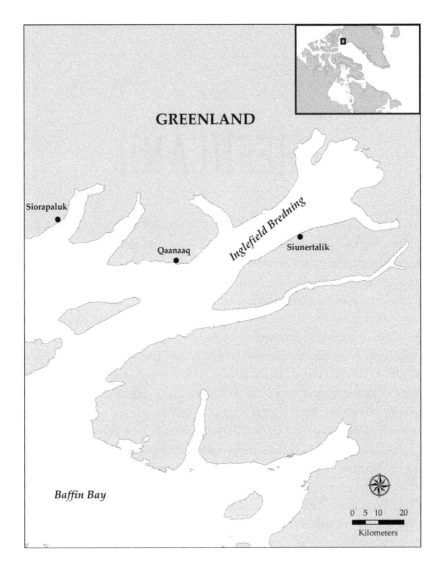

GREENLAND

Siorapaluk

Inglefield Bredning

Qaanaaq

Siunertalik

Baffin Bay

0 5 10   20

Kilometers

bergs were far in the distance, but the rest were a mishmash of iceberg rejects—dirty mounds of slush and ice like melting piles of snow left at the curb by city snowplows. I could tell that those with a crisp edge on one side were recently part of larger blocks. Others contained bluish crevasses or ice caves, and still others had muddy rocky masses sitting at water level that I thought were seals or walruses. Some had shark fins and spires, ski jumps and half pipes, and one even appeared to have a bulbous nose. Some small ones seemed to melt so fast that they totally disappeared while I watched them drift by. In the midst of it all, glaucous gulls, northern fulmars, and black-legged kittiwakes swarmed around two-man boats as fishermen maneuvered between the bergy bits and pulled up lines with substantial catches of what looked from my vantage point to be Greenland halibut and Arctic cod.

It took one more day of sitting in airports and planes before finally arriving on the windswept gravel runway in Qaanaaq, where just one commercial flight arrives each week. Despite the rainy weather, the village looked like it was laid out for a tourist brochure—every house painted in bright primary colors, teams of sled dogs tied up along the beach, small skiffs moored just offshore, the huge bay filled with icebergs of all sizes, and mountains jutting up from the shorelines far in the distance. Occasionally bits of Qaanaaq glacier could be seen peeking out from the hillside behind town. The accommodations at the five-room Hotel Qaanaaq, the only lodging available, were much nicer than the facilities in Pond Inlet, and I was the only guest for the week. The hotel living room was filled with artifacts of Arctic wildlife, including several narwhal tusks in varying sizes, a polar bear skull, three walrus skulls, a killer whale jaw, and stuffed specimens of Arctic fox, rock ptarmigan, and several seabirds.

After settling into my room, I took a stroll around town, a ten-minute walk from end to end. The village is built on a hill, and all of the houses are of just two or three designs, with older ones built near the bottom of the hill and newer ones up high. Besides homes, I saw a church, hospital, school, store, town hall, and a one-room museum. While some people are employed in these facilities, or seasonally in the factory that processes whale blubber, or muktuk, most families subsist by fishing and hunting. It appeared to be a rather young community—lots of school-age children and adults in their thirties and forties, but few elders. Those I saw smiled warmly at me, though none spoke much English. The one-page telephone directory posted on the wall of the airport and hotel listed plenty of people named Jensen, Kristiansen, Hansen, and Peterssen—clear evidence of the long-time control that Denmark has held over Greenland—but the most common name was Qujaukitsoq, a uniquely Greenlandic name.

As I continued my initial walk around town, I counted fewer than a dozen automobiles—they're mostly an unnecessary luxury in a village of that size. I was surprised to see no snowmobiles or all-terrain vehicles, which were abundant in Pond Inlet. There were, however, sled dogs aplenty—probably more dogs than people—and they howled when their neighbors were being fed. Otherwise they slept and ignored my presence. The only other noticeable sound was frequent thunder claps, which I eventually realized was the sound of icebergs breaking apart in the distance.

If I had taken a compass reading from Qaanaaq, or from almost anywhere else in the region, it would have told me that north was directly west of where I was standing. That's because a compass is a mostly useless device in high latitudes, since the magnetic north pole—the place on the surface where the Earth's magnetic field points straight down and where a compass needle points north—is constantly on the move and is rarely located at the geographic north pole or what is called "true north." Because the Earth itself is a giant magnet that is not aligned with its axis of rotation, the location of magnetic north is affected by magnetic changes in the Earth's core and the arrangement of all sorts of other magnetized rocks on the planet. Since it was first discovered in 1831, magnetic north has drifted more than 450 miles and is now located west of Canada's Ellesmere Island and traveling toward Russia at about forty miles per year.

The reason I traveled such a long way was to gain an understanding of the importance that narwhals play in Inuit subsistence and culture. To gather the complete picture of the narwhal's life cycle, health and population, I had to see for myself how they are harvested and talk to the hunters who have spent their lives studying the whale's behavior. Hunting can have a major effect on local populations of many species, and because whale hunting faces great opposition in most of the world, I felt it was vital that I hear directly from those who depend on whale meat for their nutrition and who have no other way of obtaining it except by their own hand.

So the next morning I met up with Mads Ole Kristiansen, a member of the Inuit community who makes his living as a hunter. In his late thirties and one of the only hunters in Qaanaaq who speaks a little English, Mads Ole is a happy, sincere, and congenial man who, unlike many I met from the North, seemed pleased to share what he knows about narwhal hunting with an outsider. He grew up in Qaanaaq, married, and has two children, along with eleven sled dogs and five puppies, all of which he considers part of his family. In the winter he hunts seals at their breathing holes in the ice and fishes for halibut using a long-line dropped 1,000 feet through a hole cut in the frozen sea. In March and April he travels by dog sled 200 to 250 miles north, to a site that always has open water, to hunt polar bears and walruses. In the summer he hunts narwhals.

The narwhal hunting season begins as soon as the ice breaks up and disappears from the nearby fjords, typically in late June or early July. In 2010, when I visited, the season began about three weeks prior to my mid-July arrival. With his limited English, Mads Ole explained that narwhal hunting is especially difficult early in the season when the water is still cold because narwhals have better hearing in cold water and can hear the hunters in their boats more easily. He said that his grandparents' generation established a rule many years ago requiring the use of kayaks and harpoons—rather than motor boats and rifles—for the hunting of narwhals in the local fjords, and they convinced the government to make it law, which is still in effect today, although the geographic area the rule encompasses is modified occasionally. He said the rule was designed to protect the narwhal population and demonstrate the Inuit's respect for the animals while also ensuring that narwhals would always be available to hunt in the region. Because the Inuit say that narwhals are easily frightened away by loud noises—something I observed first hand during my visits in the North—hunting from kayaks provides a quiet alternative. Equally important, narwhals are known to sink when they die, so shooting them with rifles often results in the sad waste of a valuable resource as the dead animals disappear beneath the surface before they can be retrieved. The use of a harpoon with an attached buoy or float keeps the dead whale at the surface.

Based in part on recommendations from the North Atlantic Marine Mammal Commission and the Canada-Greenland Joint Commission for the Conservation and Management of Narwhal and Beluga, hunters in Qaanaaq were allowed to kill eighty-five narwhals in 2010, a quota that the hunters wish was higher. "The government always tries to make it smaller, but we are against that," said Mads Ole. "We want more because we must survive. Narwhal keeps us warm in the winter. Too few and we won't be able to survive. We'd prefer the quota be double." And yet he said that no one ever continues to hunt after the quota has been met, and everyone always follows the requirement of using kayaks and harpoons. When I asked him how the hunters police those who break the rules, he paused a long time as if not understanding the question or the concept of how someone might not follow the rules. Eventually, he said: "That doesn't happen here. They listen. I never heard about wrong narwhal hunting here." He would know. Mads Ole is the foreman of the Qaanaaq hunting association. He speaks on behalf of the hunters when they have a beef with the government or when the government seeks information from them, and he speaks to the hunters when the government must communicate with them, like when the quota has been met for the year.

Mads Ole suggested that he take me to Siunnertalik, a hunting camp two hours by boat from Qaanaaq. We would probably stay a couple of days, he said, because the best chance of seeing narwhals is to sit and wait for them, and sometimes the wait is long. "It is sometimes good and sometimes bad because they come and go," he said. We would go the next day, he added, if the weather was good. And it was.

×   ×   ×

Joined by his cousin Paaviaaraq, we left in mid-afternoon in Mads Ole's sixteen-foot open boat, which felt like it wasn't much larger than a luxury rowboat with an outboard motor. It was crammed with hunting gear, supplies, and his handmade kayak. I sat on a musk ox pelt placed over a plastic cooler beside several hand-carved harpoons and a greasy seal bladder called an *awatuk* that I would learn later is blown up like a balloon and attached to the harpoon as a buoy so the harpooned narwhal doesn't sink. The sun came out in force just minutes before we left, and the water was calm and serene. We zoomed by an assortment of alcids—a family of small seabirds that are the northern hemisphere's version of penguins—starting with a few black guillemots and followed by small groups of thick-billed murres, which flew around us in varying directions as we darted between icebergs.

About ten minutes out, almost all the birds we saw were dovekies, which nest on cliffs by the millions north and south of Qaanaaq, first just a few in the water that took flight as we approached, then flocks of thirty or more on the surface and even more in the air. Unlike the torpedo-shaped murres or the distinctive black-and-white winged guillemots, dovekies are small with stubby bills and in flight look like black-and-white tennis balls with wings, like a variation on the Quidditch ball in the Harry Potter movies. As a flying flock turned away from us, their white bellies shined in the sun, then as they turned the other way, they disappeared from view like speeding black bullets.

As we continued on our way, the scenery was stunning. Icebergs of every shape and size glided by, some like whales' tails or aircraft carriers, others boxy or mountainous or like low flat plates, and the sun reflected off them with great intensity. At an angle some were blue, another bright white, and with the sun behind them they looked so dark they appeared more like earth than ice. The land shot up from the sea at steep angles, sometimes slightly green, sometimes tawny and rocky. The remains of glaciers flashed white on the hillsides, and some peeked over ridges hinting at their immensity beyond our view. I stared at them through my binoculars and, like author Gretel Ehrlich wrote in *This Cold Heaven: Seven Seasons in Greenland*, "my eyes darted into caves, bumped over pinnacles and fractures, traced the sensual, inward deformations of ice, the overturned fields and the weathered foliations that told the story of a glacier's life: its contortions, fractures, movements, birth, and rebirth; how it appeared to be static but wasn't; how its fins and flippers of ice were bent up by canyon walls; how, in summer, its snout, on touching sea water, lifted up, floated, and calved enormous icebergs—its life almost human."

Without warning, Paaviaaraq called out that he saw a narwhal in the distance, and Mads Ole immediately stopped the engine and came to attention. Before he even saw the whale, he was slowly preparing his kayak for launch. As he coiled ropes, inserted the razorblade dagger into the harpoon foreshaft and put on his knee-high sealskin boots, we pointed out several more narwhals. When he was ready, we helped him lower the kayak over the side, and Mads Ole quickly slid inside and was off.

The handmade kayak was constructed of a wire and wood frame wrapped in canvas and painted a light blue. Just in front of where the hunter sat was a raised platform called an *assallut*, on which Mads Ole had coiled the rope that ran from his harpoon to the *awatuk*, to keep the rope above any water that rushed over the bow as he paddled. His harpoons were just over six feet long, with self-breaking points that separate from the shaft to ensure that the harpoon tip doesn't get pulled out of the animal as it tries to escape.

Mads Ole paddled deliberately but with caution and silence so as not to alarm the whales. Within two or three minutes, he was barely visible to the naked eye. While he was in the distance searching for a narwhal to kill, dozens of them emerged right next to me and Paavi-aaraq. Five or six surfaced in rapid succession fifty yards to the south, breathing loudly, then disappearing and reappearing seconds later a few yards ahead. Another group of at least eight surfaced at once, including one old, pale gray whale and several that showed their tusks above the water line. Another two or three surfaced once close to shore and weren't seen again. From behind us came another large group. There was so much activity for about five minutes that I couldn't keep up with it, and I got confused between the sound of whales breathing and the thunder of the icebergs in the distance.

And then the activity stopped. I felt bad that all the whales were around the boat when Mads Ole was more than half a mile away, and I wanted to call out to him or whistle to alert him to their presence, but it looked like the narwhals were heading in his direction and I didn't want to frighten them. Through binoculars, Mads Ole was a tiny silhouette. Paaviaaraq and I kept our eyes and ears open for narwhals and occasionally heard one some distance away, but mostly what we heard was the rushing meltwater from nearby glaciers careening down ravines toward the fjord or the whirring wing beats of a flock of murres. An hour after he left, Mads Ole returned to the boat empty handed.

×   ×   ×

When we arrived at the Siunnertalik hunting camp, what first caught my eye was a small Quonset hut at the rear of the rocky beach that turned out to be the base of operations for biologist Kristin Laidre. I had hoped to meet her in Greenland, but our schedules were out of sync, so I visited her at her office in Seattle instead, located in the shadow of a massive bridge in a large anonymous building that serves as the home of the Polar Science Center at the University of Washington. We had spoken several times before, but I didn't know what she looked like, and when I walked in her door she immediately reminded me of my favorite modern jazz singer, Stacey Kent, with a pixie haircut and a big laugh. Laidre's early interest in marine biology had been suppressed by the time she got to high school in favor of a career as a ballet dancer. She danced professionally in Seattle, but foot injuries cut short her career after just a few years, and she decided to switch gears and return to her first passion—whales. Besides, she told a radio reporter in 2009, ballet was good training for studying narwhals. "You learn how to be miserable and suffer. You just have to keep going for the sake of what you're doing."

Laidre said that Siunnertalik is a great place to study the subpopulation of narwhals that spends its summers in Inglefield Bredning, the fjord that extends east from Qaanaaq. She typically spends three or four weeks at the camp each July with a team of local hunters in an effort to tag narwhals to learn about their migratory path.

"We started working there because we wanted to capture narwhals, but they're very hard to catch in Qaanaaq," she said. "We've tried camping at different sites on the coast, we tried living on a boat for a month and sailing around a fjord and setting nets, we've tried setting nets at Siunnertalik, and basically we haven't had any luck catching a whale anywhere. Not one. Anybody that has tried up there hasn't been able to catch one."

I could hear the frustration in her voice, the admission of failure, and we sat through a moment of uncomfortable silence as she reflected on that experience. But then she pointed out that the objective of her research wasn't to catch a whale but to answer questions about where the narwhals that summer in Inglefield Bredning go in the fall and winter. It's an important wildlife management question because in order to assure the health of the population, it's vital to know whether they are hunted in other places at other times of the year. And answering that question, she found, didn't necessarily require capturing any narwhals.

"It's what we call stock identity; identifying how groups of animals move around, and if they're hunted groups of animals then how frequently are they hunted, so you can give good scientific advice for sustainable quotas," Laidre said. "The big overarching point of working there is to get that information."

Since she had no luck catching narwhals to place satellite tags on them, she and Danish colleague Mads Peter Heide-Jørgensen looked for other ways of attaching the tags. They quickly realized that the only people who have gotten close to that population of narwhals were the hunters in their kayaks. So they designed a small transmitter modeled after an Inuit harpoon that could be thrown from a kayak and attached to the animal without having to capture the animal first. The transmitter includes a "stop plate" that ensures that the tag only penetrates an inch or two into the animal's blubber layer. Like the tags used by wildlife researchers around the world, including Pierre Richard in Canada, Laidre's harpoon tag transmits the position of the animal multiple times a day to orbiting satellites.

Laidre said that the life of a narwhal researcher is mostly spent waiting. "You just wait and wait," she said. "But it's also wonderful. Some of my best experiences have been out there. It's so beautiful and peaceful and silent, and it's a very simple life. And the company of the hunters has been great." She said that the hunters are wonderful people who work hard to feed their families and who are committed to her project, even though it means that they won't be bringing any narwhal meat home for their supper. She appreciates that their relaxed demeanor brings a sense of calm to the research, though part of that feeling comes from the lack of cell phones and email access that adds stress to everyday life back home. "We're on watch, so we're working all the time, constantly looking for whales," she said, "and hoping that some come close enough where we might have a chance to tag one." The excitement builds when the narwhals approach and when the hunters launch their kayaks and get within striking distance. During those dramatic moments, Laidre remains on shore watching from a distance, knowing that the skilled hunters will have a better chance of tagging a whale without her joining in.

The greatest number of narwhals they have tagged in any one season has been seven, and there have been some years when, despite the new device, they still have not tagged a whale. "Any kind of work with narwhals is slow," she reminds me. "You need a lot of patience, and it takes many, many years to collect data to get a sample size large enough where you can actually start to understand things."

That's assuming that the tags actually work. She said that sometimes the animals swim down to the seafloor and rub upside down on the bottom and damage or break off the tag. So even when she gets a tag on a narwhal, it's no certainty that she will actually get any useful data from it. None of the tags that have transmitted information about the location of the whales has lasted much beyond early October, a time when the whales are still lingering in the Qaanaaq area. After nearly a decade of work, Laidre has yet to learn where the animals migrate.

There are several known aggregations of narwhals in West Greenland: summering populations near Qaanaaq and Melville Bay, a fall group to the south around the town of Uummannaq, and a wintering group further south in Disko Bay on the central coast. And then there's the majority that winter amid the pack ice far offshore. It's there in the dark of winter where narwhals are most active and where so little is known about their behaviors. The one thing that is certain about their winter activities is that they spend most of their time intensively feeding, diving down to the seafloor ten or fifteen or twenty times a day to catch and eat Greenland halibut. Each of those dives takes about twenty-five minutes because they have to swim about a mile each way, the equivalent of eight times the height of the Seattle Space Needle, the tallest and most iconic structure in the Pacific Northwest, which stands just a few miles from Laidre's office.

To get a sense of the whales' habitat and how many are out there, Laidre and others, with funding from NASA, flew an aerial survey of the pack ice in April 2008 to see for themselves. Flying in a small plane from the city of Kangerlussuaq, it took two hours to reach the middle of Baffin Bay, whereupon the survey team flew eight zigzag transects back and forth over the wintering grounds—as many as they could do with the fuel available in such a small plane—before returning to the mainland. It was a follow-up to a similar effort conducted a decade earlier, and using still and video cameras and multiple observers they collected a massive amount of detailed data on sea ice conditions, narwhal numbers and locations, and habitat descriptions.

"During their migration, the ice begins to chase them south, and they arrive in their wintering grounds and the ice just forms right around them," explained Laidre. "They become enveloped by the ice. It's very dense ice, but it's moving very fast. There's a strong current that moves the ice to the south, and the ice floes are constantly changing, the leads are constantly opening and closing."

Laidre's objective in conducting the aerial survey was, in part, to quantify the amount of open water where the narwhals were found. By combining the data collected on the aerial survey with satellite images of the sea ice, she determined that just two percent of the area surveyed was open water, and there were between 17,000 and 19,000 narwhals there, or seventy-three narwhals per square kilometer of open water.

"That means you have this large density of animals that need open water to breathe packed into a very small amount of habitat," she said. "The overall habitat area is large, but what's actually usable to them is quite small."

She showed me aerial pictures of the ice, and it was easy to distinguish between ice that was several days old or more and new, thin ice that the narwhals could break through to breathe when necessary. Some of the images even captured narwhals swimming beneath this thin ice. Ground level photos showed numerous hummocks where narwhals had broken through, leaving small mounds of snow-covered debris in an otherwise flat landscape.

Looking at the pictures and listening to Laidre describe the immensely challenging environment got me wondering why the narwhals stay there, when just ten, twenty, or thirty miles farther south there is far less ice and the living conditions would be much easier, with less likelihood of a sudden freeze making it impossible to reach the surface for air. Greenland halibut must surely be found in open water as well as beneath the pack ice, right? It must be easier living in ice-free water than in the dense ice pack, so why do they remain where the risk of meeting their death in an ice entrapment is so high?

The answer, Laidre said, is partly because that's simply how they have evolved. "They really have a niche; they're totally adapted to this pack ice, more than any other northern hemisphere cetacean, and they don't have many competitors. Why go farther south when you're adapted to live in the pack ice and don't need to go farther? It's evolution. They've become adapted to being in a certain climate and exploiting it and being successful, and I think that's just what they've done."

But there's more to it than that, she added. It probably also has a great deal to do with competition and the partitioning of resources. Narwhals, belugas, and bowheads are the only whales that spend their entire lives in the Arctic, but there is a large pool of more southerly whales and marine mammals—minke and fin and humpback and blue and killer whales among them, as well as several varieties of smaller whales, porpoises, and seals—that come to the Arctic in summer to feed in its highly productive ecosystem. Those subarctic species avoid the Arctic when it's dark and ice-covered and miserably cold in the winter but move in during the spring and stay throughout the summer and early fall. The narwhal has developed a strategy to exploit the eco-system at a time when there are few competitors in an area where they know they have a reliable food supply available. Belugas and bowheads do the same thing—they feed intensively in the winter and early spring when the other subarctic species aren't there to compete with them. When the ice recedes, the Arctic whales move north just as slews of subarctic species arrive in the area they just abandoned.

As for the theory that the narwhals live in the pack ice to avoid killer whales, Laidre is skeptical. "I don't necessarily think that's the case, because there are basically no killer whales in the Arctic in the winter. There's no good reason to just hide in the ice when, at that time, there is nothing to hide from. I think they go to their offshore winter-ing grounds because they offer highly predictable food resources and, to some extent, they are also partitioning the ecosystem with other species."

That's not to say, of course, that wintering in small leads amid dense pack ice isn't dangerous. Ice entrapments, called *savssats* by the Inuit, are an increasing concern and they probably happen much more often than records suggest, since the region is so thinly populated and no one is around to observe or document entrapments in the vast major-ity of the region. Nonetheless, Laidre noted that four ice entrapments that resulted in the deaths of more than 700 narwhals occurred in 2008 and 2009—the first one ever documented in East Greenland, as well as two in the Qaanaaq area and a very large one near Pond Inlet.

Laidre is beginning to examine the distribution and timing of known ice entrapments and look at the trends in the breakup of sea ice on the narwhal's summering grounds. She has found what she calls "strongly significant trends" that suggest that the ice is forming later and later. "Over a thirty-year period there is a three- to- four-week dif-ference in when the ice forms," she said. "If ice formation is a clue to the narwhals that it's time to get out of their summering grounds, then the trigger is changing, the pattern is changing." Is that change in the for-mation of ice making narwhals more vulnerable to ice entrapments? Laidre hypothesizes that it may be the case, though there is precious little data from which to draw conclusions just yet. But the hints she has found so far are another indication of the dangerous implications of global warming.

EIGHT

SUBSISTENCE

THE NARWHAL HUNTING CAMP OF SIUNNERTALIK, LOCATED half way along the Inglefield Bredning fjord east of Qaanaaq, is a flat pebble beach about 100 yards long, where, in addition to Kristin Laidre's Quonset hut, two tents were in place for hunters when I arrived with Mads Ole. Dominating the beach were scattered narwhal skulls and vertebrae and the decomposing carcasses of two Greenland sharks that had been caught earlier in the week after they tried to feed on a narwhal that the hunters had just killed. It wasn't a pretty sight, but with the exception of the lack of trees, a hunting camp in Maine or Montana probably doesn't look much different.

As we unloaded the gear from the boat, I noticed two people atop a boulder ridge on the western edge of the camp, spying through their binoculars for narwhals. It was Gedion Kristiansen and his wife, hunters who had arrived earlier in the day. We joined them to scan the fjord for narwhals, and within minutes we saw a small group of whales in the distance. Without a word, Mads Ole and Gedion jumped into their identical kayaks and paddled out to intercept the narwhals. The animals surfaced once more, but disappeared. The hunters remained in their kayaks for about an hour before returning to camp, hoping that more whales would come by, while the rest of us watched for additional narwhals from the observation point.

It wasn't long before Mads Ole's earlier comment became clear to me. He said that narwhal hunting is mostly about waiting: sit high on an overlook in the cold and wind and stare through binoculars while waiting for a narwhal to surface within paddling distance. That could take hours or days, depending on the weather, the currents, and the whim of the narwhals. When a narwhal is sighted, the hunters rush to launch their kayaks, but even then they spend a long time sitting quietly on the water waiting for the animals to resurface, then reposition themselves and wait again for them to come closer. When the whales don't come close enough, the hunters return to camp where the waiting and watching resumes. For the next ten hours, we waited and watched. I was cold within the first fifteen minutes, which made me worry about how I was going to make it through the next two days while also impressing upon me how easily the hunters and their families endure a lifetime of such challenging conditions, and much worse.

I was also nervous. I was in a place I knew nothing about, where the culture and language was unknown to me, among people with whom I could barely communicate, but upon whom I was totally dependent for my survival. I have spent plenty of time in wildlife research camps with biologists studying an amazing variety of animals, but this strange environment, as beautiful and peaceful as it was, forced me to remain constantly on high alert. Besides, I had never been hunting before, so I was uncertain of even the simplest protocol and what the hunters expected of me.

Time has little meaning at an Arctic hunting camp. At 3 a.m., five hours after I would normally have been asleep, the camp was still lively, with people up eating and drinking tea. Another family of hunters had arrived, their two children were still awake and exploring the nooks and crannies around camp, while the men sat on a driftwood log talking quietly and watching for narwhals. Their wives watched from the observation point. In the twenty-four-hour daylight, I was still mostly awake, too. Even the gulls and fulmars soared about, cackling nearby like hens encouraging each other to produce another egg. I concentrated on scanning for narwhals and watching the icebergs, which Mads Ole said were calved from one of seven glaciers nearby and which, because of the unusual currents, travel out of the fjord along the far shore and circle back along the near side before circling again and departing the area. "Maybe next it goes to your country," he said laughing.

Mads Ole and I were warming up in the tent when, at 4:02 a.m., Paaviaaraq called out from the lookout that narwhals were approaching. Before I even understood what was happening, Mads Ole dashed out of the tent, ran up to the observation point, and quickly verified the location of the whales. He jumped into his kayak and took off, followed shortly by Gedion and soon after by the third hunter, Thomas

Qujaukitsoq. When I finally climbed to the lookout and raised my binoculars, I could see dozens of narwhals in all directions—some in groups of two or three, others in family groups of a dozen or more, and all apparently exiting the fjord together. From my vantage point it appeared that the hunters were surrounded by narwhals, some of which looked as if they were very close to the kayaks, yet the hunters sat still, not paddling closer, and never hoisting their harpoons. It was exciting to watch, but probably quite frustrating for the patient hunters. Mads Ole told me later that the wind had come up in the night, which made for poor water conditions for hunting. He wasn't optimistic when he climbed into his kayak because the waves were too high, and he never was able to get close enough to a narwhal to throw his harpoon. An hour later, the three hunters returned to camp, but as they climbed out of their kayaks, another call came that more narwhals were approaching. Only Mads Ole decided to try again.

Within minutes he was in perfect position to intercept at least a dozen very dark narwhals swimming in a line toward him. He waited, unmoving, took a couple of paddle strokes, then waited some more. The narwhals came closer, and several swam right by him, appearing to be just an arm's length away, and still he didn't move. Three narwhals surfaced at once just behind him, and as they went under he paddled alongside them and lifted his harpoon. But they resurfaced too far away, so he lowered it again. I thought he had missed his best opportunity, as I didn't see any more narwhals in the vicinity, but two whales trailed the group, surfaced, dived, and when they surfaced again, one was right beside Mads Ole. He quickly raised his harpoon, held it in the air for what I thought was far too long, and threw it. The harpoon struck the narwhal in the flank, and the animal responded with a sharp slap of its tail and dived, taking with it the sealskin float.

A celebratory cheer rang out from the hunters and their families, and from me as well, and everyone rushed into action. Mads Ole called out for another harpoon and *awatuk*, and we all jumped into one of two motor boats and raced towards him. In the minute it took to reach him, I noticed that the float wasn't visible. The narwhal had taken it down with her. In another minute, the extra harpoon was handed to Mads Ole, and another cheer went out as the float popped to the surface, still attached to the whale. As the injured narwhal repeatedly surfaced for air, clearly in distress, with blood pouring from its wound,

Mads Ole struck it with the second harpoon and float, ensuring that the animal wouldn't sink. Within seconds, Gedion pulled a rifle from a compartment in the boat I was in, fired a single shot, and the narwhal lay dead at the surface. The hunters quickly tied its tail to the bow of one boat, retrieved the harpoon tip from the whale's flesh, and the entire assemblage returned to camp.

×   ×   ×

Whales have been hunted in the Arctic for thousands of years. The hunting of bowheads, the largest of the whales that call the icy waters of the North their year-round home, may have even helped the early Inuit to expand from the western Arctic to eastern Canada and Greenland. Due to their relative abundance, the much smaller belugas have long been the most commonly hunted whale in much of Canada. In Greenland, narwhal ivory had been traded among the Inuit long before the culture had its first contact with Europeans. Narwhal tusks were used whenever they needed something straight and hard, such as tent poles, tools, and harpoon foreshafts; the sinews for sewing waterproof boots and clothing; the oil for lighting; and the meat and muktuk—the top layer of skin and blubber—were eaten by both people and dogs.

The commercial hunting of whales in the Arctic began in the late 1700s when whaling ships from various European nations came in search of bowheads. Their annual trips to the region took them into Lancaster Sound, north of Baffin Island, and eventually into Hudson Bay. In southern Greenland, humpback whales were also targeted. Wherever the whalers went, Inuit were hired on as crew on the wooden whaleboats, and they also transported baleen and blubber by dog sled from the edge of the ice. When bowhead numbers declined significantly, whalers focused more on hunting narwhals for their ivory and their superior oil, which was used for heating and lighting.

When commercial whaling from European and American boats ceased, Greenlanders continued hunting whales, using a 127-ton Danish vessel between 1924 and 1949 that could catch even the fastest swimming whales. In 1927, for instance, it reportedly caught twenty-two fin whales, nine humpbacks, seven blue whales, and two sperm whales. As that boat was phased out, Greenlanders revitalized their community-based whaling when some fishermen installed harpoon cannons on their boats. Inuit hunting of minke whales in southern Greenland continues today, while narwhals are the predominant whale hunted in the North.

In Canada, where twenty-six native communities are legally permitted to harvest narwhals, hunting is conducted from the floe edge in June, in ice-cracks in mid-summer, and in open water in August and September. Strict quotas for each community are maintained—all are between five and fifty animals per year, except for Pond Inlet (130), Arctic Bay (130), Qikiqtarjuaq (90) and Repulse Bay (72). With the exception of Arctic Bay, the communities have seldom come close to their harvest limits in recent years. Canadian hunters aren't restricted by law, as the Qaanaaq hunters are, to using only harpoons and kayaks during the open water season, which most Canadian communities abandoned in the 1950s. In fact, the Nunavut Land Claims Agreement says that the Canadian government cannot place restrictions on how, where, or when the Inuit may hunt. Instead, the agreement states that the Nunavut Wildlife Management Board is "the main instrument of wildlife management" and "the main regulator of access to wildlife" in the region. The board makes decisions about quotas and related issues, and the local hunters and trappers organization in each community decides what gear to use and when the hunting season begins and ends. Hunters in almost every community use rifles and motor boats to hunt narwhals in the open water, although they do use harpoons on occasion to attach floats to dead whales to prevent them from sinking, or hand-thrown grappling hooks to retrieve them.

The tusk, which can sell for about $150 per foot, is clearly the most economically valuable part of the narwhal to Inuit hunters. But an export ban in Greenland since 2004 and a similar ban in seventeen hunting communities in Canada announced at the end of 2010 (partially revoked in late 2011) make it difficult for the hunters to earn their full value, since foreigners usually pay the highest price for them. In both cases, the bans were instituted because government scientists could not issue a statement of "non-detrimental findings" certifying that hunting in those areas would not harm the narwhal population. When new information is gathered about the narwhal population, the Greenland ban is expected to be repealed. Canadian officials restricted the export of tusks in part to stop the international community from banning exports altogether under the Convention on the International

Trade in Endangered Species, a treaty that protects wildlife from over-exploitation. The hunting communities in Nunavut appealed that decision, and a negotiated agreement in 2012 restored hunting to most of the villages.

Besides the tusk, the most prized part of the narwhal as a source of food in both Greenland and Canada is the muktuk (spelled variously as mattak, maktaaq, maktaq, and mungtuk in different Arctic regions). It's a delicacy rich in vitamin C that is most often eaten fresh and raw and remains a staple in the traditional diet. In many Inuit communities, especially in Canada, the meat of the narwhal is primarily used as dog food.

Commercial whale hunting receives considerable support in only a few nations, notably Japan, and even there it faces significant opposition. But there is less opposition to subsistence hunting of whales by aboriginal populations who have hunted the animals for centuries, even in the United States, though the issue is by no means resolved. Hunting, wrote Barry Lopez, "is the most spectacular and succinct expression of the Eskimo's relationship with the land, yet one of the most perplexing and disturbing for the outsider to consider." In *Inuit, Whaling, and Sustainability*, written in 1998 to make the case for continued whale hunting by the Inuit, author Milton Freeman and colleagues defined the ongoing disagreement as primarily

> a cultural conflict, one between different sets of values, between different philosophies and ways of understanding and appreciating nature and its wondrous creatures. It is the conflict between, on the one hand, a hunting people living for the most part in small coastal hamlets and towns in the Arctic, and, on the other hand, urban peoples living outside of the Arctic who, for a variety of reasons, now believe that whales should not be killed for food. However, despite our need to eat whales and other animals we hunt, we also share city dwellers' fascination and respect for these magnificent animals.

Inuit communities have a mixed economy that combines the traditional sharing of food among community members with the buying and selling of meat and other products. The factory in Qaanaaq—where the local whale hunters sell about seventy percent of the muktuk from each narwhal they kill, while keeping the rest for their family and community—packages the fresh muktuk for sale to cities in southern Greenland where residents have little access to it but wish to maintain a traditional diet. The money made from the sale of muktuk and narwhal tusks is the only currency many hunters earn to pay for clothing, food, fuel, and other necessities.

Whaling also helps the Inuit maintain a link to their cultural heritage. It is a common theme in legends, songs, dances, and art, and while many ceremonies and rituals are no longer practiced, whale hunting allows members of the community to pass on to future generations some of the distinctive elements of their culture. "The widespread sharing [of muktuk] among relatives and between communities creates and sustains the bonds that remain the basis of Inuit social and economic relations in the North today. These relationships are indeed critical to Inuit cultural survival," wrote Freeman.

Perhaps most important, the consumption of muktuk is vital to the health of Arctic residents, who have little access to fresh fruits and vegetables during most of the year and who, without whale products, are left with unhealthy processed foods shipped in from thousands of miles away. Fats derived from marine animals protect against cardiovascular disease, and scientists have attributed the almost complete absence of heart disease in Arctic communities to the fatty acids they consume from the oils in whale products. Muktuk is also a major source of antioxidants and selenium, which protect against the harmful effects of mercury often found in unsafe levels in foods throughout the Arctic. Compared to store-bought beef, pork, or chicken, whale products provide superior nutritional value.

As Mads Ole told me, "If I live here without whale meat, I'm nothing, because nature is very strong. In winter, if I don't have meat to eat, I'm always cold. If I mostly eat vegetables, I become cold, because they don't keep you warm."

These arguments in support of subsistence whaling don't sway many people to the south, for whom the killing of whales for any purpose is unacceptable. The Canadian Marine Environment Protection Society, for instance, opposes all whale hunting, whether commercial or subsistence. Part of their opposition is due to their disagreement with the official government population estimates, which the society believes are exaggerated. "The government's higher population numbers seem to be largely based on traditional knowledge. Independent researchers believe that the reason more Inuit are reporting seeing more whales than ever is because of the rapid growth in the human population of Nunavut since becoming a province," wrote the group's vice president Annelise Sorg in an email to me. "As well, the introduction of motorized land and ocean vehicles equipped with radar and other electronic gadgets help the modern Inuit travel faster and farther and therefore spot more whales."

Furthermore, she believes that whale hunting quotas in Canada are set too high and that the quotas are routinely surpassed without any repercussions to the hunters. Sorg argues that many of her concerns could be more easily addressed if the Canadian government would rejoin the International Whaling Commission, the body that oversees commercial whaling throughout the world. Canada withdrew from the group in 1986 after the commission passed a moratorium on whale hunting, claiming that the ban "was inconsistent with measures that had just been adopted by the IWC that were designed to allow harvests of stocks at safe levels."

Sorg and the Canadian Marine Environment Protection Society, along with the Whale and Dolphin Conservation Society in Britain and others, endorse the Declaration of Rights for Cetaceans, which was written following a 2010 meeting of philosophers, scientists, and conservationists in Helsinki. It states, in part: "Based on the principle of the equal treatment of all persons; recognizing that scientific research gives us deeper insights into the complexities of cetacean minds, societies and cultures; noting that the progressive development of international law manifests an entitlement to life by cetaceans; we affirm that all cetaceans as persons have the right to life, liberty and wellbeing."

That's not the position taken by independent biologist James Finley, though. He argues against whale hunting not on moral or ethical grounds but on evolutionary grounds. Finley supports subsistence hunting in Inuit communities, but he strongly opposes narwhal hunting for the primary purpose of obtaining and selling their tusks. "The use of the gun to eliminate any skills necessary to procure meat, and our insane big game hunting culture, which has macho males strongly selecting for big bucks, has had a profound and very rapid impact on any species that carries big antlers, teeth, or horns," he told me via email. Finley said that, from his experience, native hunters would never kill a male narwhal in its prime, largely because their meat is tough and nearly inedible, unless it was being killed for its tusk. "Our crazy culture, which is increasingly emulated by younger Eskimos, thinks our prowess is best measured by the size of the rack of horns over the fireplace," he said. His greatest concern is the effect that harvesting the whales with the largest tusks will have on the evolution of the animals. He argues that harvesting only the oldest and most ornamented whales means that those who carry the best genes for survival will face the greatest hunting pressure, meaning those with inferior genes are more likely to live longer and reproduce. He points to several recent studies, especially one on bighorn sheep in the Canadian Rockies, which have shown that such strong selection pressure has had a profound genetic effect by reducing the size of the main sexual attractant. It also can have profound effects on social organization and breeding success.

Finley also takes issue with the selling of muktuk, in large part because, like the bush meat trade in Africa, putting a price on wild meat can have serious ecological implications when abundant human populations seek to kill potentially vulnerable animals to make money. "A price on wild meat often puts populations in jeopardy," he said.

×   ×   ×

Finley's point about the younger generation of Inuit hunters—distanced from the semi-nomadic life of their ancestors and interested less in respectfully harvesting only the animals they need and more in the cash they can make from selling a tusk—is echoed elsewhere. Alaskans have struggled in recent years with young members of their northern communities killing walruses and cutting off their heads to sell their tusks while leaving the rest of the animals to rot. It is apparently an issue in the eastern Canadian Arctic as well, though I found no one willing to talk about it. In fact, I could find almost no one who would talk about narwhal hunting in Canada at all. The government scientists only hinted at the difficult politics involved, the wildlife managers wouldn't return my messages or wouldn't go on record as saying anything substantive, and the hunters themselves, including the local hunters associations, didn't want to have anything to do with me.

Much of the Inuit's reaction can be at least partly attributed to an article in *National Geographic* magazine in 2007 by Paul Nicklen, who grew up on Baffin Island and is one of the foremost photographers of

Arctic marine mammals. His story about a narwhal hunt he observed in Canada, illustrated with graphic and often disturbing photographs, suggests that hunting practices may need to be reviewed and record-keeping expanded. At the very least, it suggests that wildlife managers should be paying much closer attention to narwhal hunting because the number of animals recorded as being harvested may be far lower than the number actually killed, since many appear to be what they call "struck and lost."

Nicklen wrote that during one twelve-hour span, he counted 109 rifle shots but just nine narwhals were recovered. One hunter reported that he killed seven narwhals, all of which sank. "This was not the first time I had heard reports of many narwhals being shot but few landed. Just weeks earlier, a man I know to be a skillful hunter confided that he had killed fourteen narwhals the previous year but managed to land only one. . . . So much ivory rests on the seafloor, said one hunter, that a salvager could make a fortune," wrote Nicklen.

My only direct experience with narwhal hunting in Canada occurred during my first visit to Pond Inlet when my group left the observation point at Bruce Head because several young hunters had arrived. In the next thirty minutes or so, while we drifted in the water half a mile away watching small groups of narwhals surfacing around us, we heard at least nine gunshots. Later, we saw the remains of one narwhal on shore, its tusk removed and only a small quantity of its muktuk sliced off. The hunters reported that they had killed one more narwhal but it sank before they could retrieve it. If that is the typical outcome of a narwhal hunt on Baffin Island, and Nicklen's article suggests that it may be, then I certainly don't come away with the positive feeling I had after observing the hunt in Greenland. Young gunslingers using whales for target practice in hopes of landing one of several that they kill, focused entirely on the holy grail—the money they can make from a long tusk—while making little use of the rest of the animal, is a practice that I find extremely troubling. It's no wonder that the community doesn't want to talk to outsiders about it.

Yet I wish they had. I wanted to give them a chance to respond to Nicklen and Finley and demonstrate how Canadian narwhal hunting practices of today are just as respectful and dignified and appropriate as what I experienced in Greenland. I was hoping that someone would say that Nicklen's article was hurtful and inaccurate or that my observation of the young hunters was an unfortunate anomaly. It may be. But I could find no one who would defend them in even a cursory way.

While the Canadian Inuit were, by all accounts, angry with how they were portrayed in Nicklen's article, they were also apparently uncomfortable with the global media attention drawn to their harvest of more than 600 narwhals trapped in ice in November 2008. The ice near Pond Inlet apparently closed suddenly, leaving the whales just a few small breathing holes, and those were closing, too, while the next nearest place to surface and breathe was thirty miles away. The animals were certain to die, so the Pond Inlet Hunters and Trappers Organization advised its members, with approval by federal officials from the Department of Fisheries and Oceans, to kill the narwhals rather than let them suffer. In this case, the harvest seemed entirely appropriate to me—the entrapment was a natural event, the whales were certain to die after what would surely have been a long and stressful experience, and their harvest by the hunters not only reduced the animals' suffering but provided an unexpected bounty to the local Inuit communities. Tissue samples from many of these entrapped narwhals were studied by government scientists, and an alarmingly high number were reported to be infected with brucellosis, a bacterial disease that can cause weight loss, infertility, and lameness. While it is unlikely that the disease contributed to the entrapment, it raises concerns about the health of the narwhal population and the increased risk that brucellosis could be spread to people who eat narwhal flesh.

Entrapments such as this are extremely rare. David Angnetsiak, a fiftyish hunter in Pond Inlet, told me it was the only time in his life that he had seen narwhals inescapably trapped. He said that during the entrapment there were about ten or twelve breathing holes in close proximity, each about ten to fifteen feet in diameter, and there was no chance that the holes would remain open throughout the winter. He said it took two weeks of constant work during the five daylight hours each day—about 9 a.m. to 2 p.m.—for the hunters to shoot and retrieve all of the entrapped narwhals. The hunters worked in rotating teams, with one hunter shooting, another following up with a harpoon, others tying a rope around the flippers, and still others hauling the animals up on the ice and flensing them. The entire Pond Inlet hunting community was involved, with no hunters from outside the region allowed to participate, though muktuk was shared in many Arctic communities.

Angnetsiak said that most of the entrapped narwhals were females and young, noting the Inuit belief that bull narwhals can hold their breath longer and were probably better able to reach open ocean.

While news reports indicated that federal officials said the harvest was the most humane way to deal with the entrapment, there was considerable public outcry that the government didn't send in an icebreaker to open a passage for the trapped whales. The Humane Society International/Canada called the government decision "unconscionable" and noted that the narwhal harvest was "inherently inhumane." Others from the South agreed, though Mike Richards, a special administrative officer for Pond Inlet, told the *Globe and Mail* of Toronto that "It's just a misfortune of nature." The last such entrapment that local hunters remember occurred in 1943.

Mads Ole Kristiansen told me that ice entrapments of narwhals happen very seldom in Greenland, too, though given the sparse population in the North it's difficult to know for sure. Records indicate that the largest entrapment in Greenland occurred in 1915, when more than 1,000 narwhals died, and in 1943 about 350 narwhals were harvested from an entrapment. Mads Ole believes that global warming may be causing more frequent entrapments, noting that after the waters around Qaanaaq froze over in 2009, a severe storm broke up the ice, narwhals returned to the area, and then the water quickly froze again and trapped a small number of narwhals, which were killed by local hunters.

He may be on to something. Researchers from the Arctic Institute of North America reported in 2009 that changing sea ice patterns due to climate changes may have led to significant increases in narwhal catch rates around the village of Siorapaluk, just north of Qaanaaq. Narwhal catch numbers there experienced a sharp increase that researcher Martin Nielsen said could not be explained by changes in hunting effort. Instead, he said that hunters attributed the increase to changing sea ice conditions in Smith Sound, between Greenland and Ellesmere Island. The hunters told him that sea ice was forming later and breaking up earlier, giving the hunters greater access to narwhal stocks in the area.

It's a phenomenon that scientists and Inuit hunters will be paying close attention to in coming years.

NINE

MUKTUK

GREENLAND IS A LAND OF CONTRASTS, OF DARKNESS AND light, of modern and traditional, of ice and rock, of life and death. Its landscape, as Gretel Ehrlich wrote, "with its shifting and melting ice, its mirages, glaciers and drifting icebergs, is less a description of desolation than an ode to the beauty of impermanence." With a coastline almost as long as the circumference of the Earth, it's a nation that only recently won a sort of independence when in 2009 it gained self rule from Denmark, enabling it to make political decisions that previously had been decided a thousand miles away. While Denmark still controls foreign affairs, currency, and the justice system—as it has since the eighteenth century—the official language of the world's largest island is now Greenlandic and the island government and its residents now hold the rights to its minerals and other natural resources. But it was a long time coming.

Greenland was populated as a result of a series of migrations over the course of about 4,500 years, but the island has also experienced long periods when it was virtually uninhabited because environmental conditions made it too harsh to survive or there were too few animals to hunt. The first wave of immigration brought the Saqqaq people from Canada to the coasts of central and southern Greenland between 2500 B.C. and 900 B.C. Four hundred years later and lasting for about 500 years, the Dorset culture made its way across the ice from Canada to northwestern Greenland and spread southward as land-based hunters of reindeer and musk ox. During most of the first millennium A.D., the country is believed to have been uninhabited, but the third wave of immigration between the eighth and tenth centuries brought three different cultures to the shores of Greenland: a second group of Dorset people from Canada to northern Greenland; the Thule people, also from Canada, who settled the east and west coasts as maritime hunters and fishermen; and Norse settlers from Iceland and Norway, led by Erik the Red, who arrived in South Greenland in 982 A.D. Primarily farmers, the Norse disappeared from Greenland in the early 1400s, with the last account of their existence coming from a description in the Vatican's annals of a wedding held in Hvalsey Church in 1408.

Today's native Greenlanders are descended from the Thule people, whose name comes from the mythical land that was thought to be the most northerly place on Earth. Greek explorer Pytheas was the first to claim to have seen Thule when he sailed for six days north of Scotland in the fourth century B.C. Pliny the Elder described it as "the most remote of all those lands recorded," and every time an explorer made a new discovery to the north, it was called Thule. The Romans claimed they had conquered it. Christopher Columbus said he was there before arriving in America, as did many others, including ethnographer Knud Rasmussen, perhaps the most famous Greenlander of the last century, who traveled the region collecting myths and legends from the native people.

Rasmussen's father, born in 1879 in Jacobshavn, was a Danish missionary who married an Inuit woman, and Rasmussen's first language was Greenlandic. In 1910 he created the Arctic Station of Thule, the northernmost outpost in the world just south of present day Qaanaaq, and used it as a base for a series of expeditions to study native people and customs while also mapping the coast of northern Greenland. His fifth expedition, a four-year trip that began in 1921, is perhaps best known, as he became the first person to cross the Northwest Passage by dogsled while learning about the folklore and culture of the people he met along the way. The book containing the treasury of stories he collected, *Greenlandic Myths and Legends*, continues to be the primary source for those interested in Arctic folklore. The house Rasmussen lived in while in Thule is now home to the historical museum in Qaanaaq, where I received my introduction to the man.

But why the house traveled the nineteen miles to Qaanaaq is also a story worth telling.

Thule Air Base is the northernmost installation of the United States Armed Forces. It is home to a global network of sensors providing missile warning, space surveillance, and space control to the U.S. military. It was constructed in secret during the early 1950s, and in 1953 the residents of the nearby town of Thule were given four days notice to move north to establish a new community, now called Qaanaaq. Hans Jensen, who owns the Hotel Qaanaaq where I stayed, was two years old when everyone was forced to move, and while he has no memory of it, the stories told by family members suggest it was a tremendously stressful and traumatic experience. There were no houses in place yet, so everyone was forced to live in freezing tents until the government provided homes sometime later. Eventually, Rasmussen's house was moved to its present location, and a museum was established in it that showcases artifacts from the Dorset and Thule cultures and exhibits about Rasmussen's explorations of the region.

×   ×   ×

While there are still many unanswered questions about narwhal biology and ecology, Kristin Laidre has found a new and unusual use for what may be the narwhal's greatest behavioral claim to fame—its ability to dive to great depths. In 2006 and 2007, Laidre successfully attached satellite transmitters to fourteen narwhals in Melville Bay. In addition to collecting data on the animals' geographic positioning, the tags also recorded the temperature of the water as they dived to the bottom. While Laidre knew that temperature data would be somewhat insightful to her narwhal research, she also guessed that oceanographers and climatologists might find the data useful as well. It is difficult to use traditional oceanographic measuring devices to monitor water temperatures at great depths, especially in the winter when the water is frozen over for many months. As a result, researchers have little data from the middle of Baffin Bay to feed into their climate models. Instead, they have used temperature readings from coastal locations or estimates calculated from historic data, which clearly provide only a hint of the true picture.

"The gliders that oceanographers occasionally use to collect this data sort of look like a streamlined narwhal, with a long pointy antenna, but they're not as smart as narwhals because sometimes they can't find the surface and they get caught under the ice," Laidre said. When the narwhal-collected temperature data was published in a scientific journal in 2010, it generated considerable media attention, not only for what oceanographers learned about water temperatures but also for the unusual way the data were collected. Thanks to the narwhals, water temperatures were determined to be about 1.8°C warmer than what the climate models predicted, and the surface isothermal layer—a layer of water that is at a constant temperature—was 160 to 260 feet thinner than previously believed.

"I'm not an oceanographer, but for me the excitement was the proof of concept, that it worked and we collected useful data and the scientific community seemed to be interested in it," Laidre added. "[The data] suggested that Baffin Bay has continued its warming trend, though I don't know that it was terribly surprising to the oceanographers."

As fun a project as it was to use narwhals as thermometers to take the temperature of the water in the middle of Baffin Bay, it didn't begin to answer the questions that Laidre still yearns to answer—starting with how many narwhals live along the Greenland coast and what route do they take on their migrations. She said that a 2007 estimate placed the summering population in the Qaanaaq area at around 8,000 animals, with another 6,000 summering in Melville Bay. Of those animals, Laidre said that no one knows for sure where the approximately 7,800 narwhals that spend the winter in Disko Bay come from. Are they from among the group that summers in Qaanaaq or do they come from other areas? Probably both. But she knows that some narwhals that summer in the Canadian archipelago find their way to Disko Bay as well.

The good news is that narwhal numbers in West Greenland are rebounding after a long decline caused by overhunting. Records of narwhals killed in the waters of Greenland have been maintained since 1862, and they show that about 200 animals were harvested on average each year until about 1960 when they slowly increased to between 600 and 800 in the 1980s, with one year topping out at 1,400 narwhals killed. There was a steady decline beginning in 1990, primarily because narwhals were scarce in the region. Concerns expressed by the scientific community about the health of the narwhal population finally forced the government to implement hunting quotas in 2004, with 300 narwhals allowed to be harvested each year in West Greenland. Quotas were set for East Greenland for the first time in 2009, where narwhal hunting records date only to 1955 (with several years of missing data) and where narwhal harvest levels in the villages of Ittoqqortoormiit and Tasiilaq had increased significantly until recently.

In her 2006 book, *Greenland's Winter Whales*, co-written with Mads Peter Heide-Jørgensen before the hunting quotas were established, Laidre wrote that narwhal populations in Greenland were at their lowest levels since the last ice age. "This is undoubtedly because of the large harvest over the past twenty to thirty years. The growing human population, the use of motorized boats, and the increasing value of the much demanded mattak (muktuk) are partial explanations for the harvest levels." In 2010, however, she told me that the quotas resolved the overhunting issue in Greenland, and the annual harvest has been reduced by hundreds of whales, enabling the population to begin to recover.

When pressed about the issue of narwhal hunting, Laidre pointed out several differences between the situation in Canada and Greenland. She said there is a very different process for how quotas are set in the two countries, with the government establishing quotas for hunting communities in Greenland whereas the communities themselves determine their own quotas in Canada. She also noted the kayak-and-harpoon hunting technique required in Inglefield Bredning and Melville Bay, though motorized boats and firearms are allowed in more southern communities in West Greenland, just as is the case throughout the narwhal range in Canada.

"I think the real distinction is a hunt driven primarily by a tusk [in Canada] and a hunt driven for subsistence for food [in Greenland]," she said. "There's no doubt that Greenland is changing and becoming more globalized, but if you choose a hunter's life, it's not a rich life. You provide for your family and your dogs and your extended family." The number of full time registered hunters is declining in Greenland, especially among men in their twenties and thirties. In previous generations, that's how old most of the hunters were, because they were typically the strongest and toughest men who would choose to engage in the rigors of the hunting life. But as those hunters age, they are not being replaced by a new generation of hunters because Greenlandic youth are becoming enamored of the Western lifestyle. "Do you choose the life of a subsistence hunter and keep these traditions going," asked Laidre, "or do you get an education, which means you leave your community and leave everything around you and expand your horizons?"

×   ×   ×

When we were back on shore at the Siunnertalik hunting camp after the successful hunt, Mads Ole judged that the narwhal he had killed that day was a four- or five-year-old female. He said that, from a kayak, it's usually impossible to determine whether he is throwing his harpoon at a male or female, so it is much more difficult to target males with tusks. This one—the first I had ever been so close to—was about nine feet long with beautiful mottling of grays and tans throughout its flanks and an attractive starburst of white at its navel. It had several whitish scratches on its skin, tiny eyes, and a small mouth that appeared to be curved, like a beluga's, into a perpetual smile. Its body was soft to the touch, like a firm steak or an inflated inner tube, and its melon (the rounded part of their head containing blubber oils that they can alter in shape) felt even softer still.

Mads Ole and Paaviaaraq began to flense the whale. With several long, sharp knives used exclusively for this purpose, the men made a straight slice down the narwhal's back from head to tail, causing a large quantity of blood to ooze out. They removed the animal's flippers and cut off eighteen-inch square sheets of muktuk from around its body, tossing each slab into the gently lapping salt water to be cleaned and cured. After cutting off the tail, they removed the inner layer of blubber—it looked like pink foam insulation, but it felt like uncooked fat—and fed it to the flock of begging seabirds. Next, they carved huge black steaks from the animal's muscle, and cut out its spinal column for the dogs to chew. They carefully removed the internal organs and stacked them on the beach, starting with the lungs, liver, and heart, which Mads Ole said tastes quite good. The large intestines were coiled like a hose. When I asked what he does with the intestines, Mads Ole smiled and said, "I don't know, but it's on my wife's wish list."

After cutting open the narwhal's stomach and seeing that its last meal was a modest number of small shrimp-like creatures, Mads Ole tossed it into the water with the remaining organs to feed the sharks, though the fulmars at the shore had first dibs. Once the hunters cut up the ribcage, all that was left was the head, and even that was sliced up for dog food. In the end, only about five percent of the whale was discarded. The rest would be consumed by the local community and their dogs, with the exception of some muktuk that would be sold to the factory for distribution elsewhere.

I was totally surprised at how little negative emotion this experience generated in me. Without even thinking about it, I had helped the hunters point out narwhals in the water, was excited for them as they neared a kill, and cheered when it happened. I was right next to the gun that ended the narwhal's life—the first time I had ever been within 100 yards of a gunshot, and it was jarring—and yet I was somewhat detached. Intellectually, I've always opposed sport hunting but supported the idea of subsistence hunting. And now, having witnessed a subsistence hunt, I could see its value and not be put off by the experience.

Yet I'm still surprised at myself. I've been an animal lover and wildlife conservationist my entire life; I cringe when I see sporting shows on television where a turkey or deer is killed; and I am aghast when I run over a chipmunk or when a bird strikes my window and dies. I am still tormented from a moment as a teen when a neighbor swerved his car to intentionally strike a raccoon, which were considered pests in our gardens. I don't understand why anyone would do such a thing. Yet my narwhal hunting experience didn't seem to bother me at all, not even the flensing, which I thought would disgust me. I've never seen my food killed—except lobsters and shellfish going into a pot— nor have I seen a large animal autopsied, but the flensing looked simply like a classroom dissection or science experiment. Maybe I've observed enough wildlife research projects and talked to enough biologists who have taken plenty of lives in the name of conservation that I've become desensitized to those feelings. If that's true, it's news to me, since just anticipating this experience had me nervous and uncertain whether I could stomach it. Apparently I can.

×　×　×

For the rest of the day at Siunnertalik, few people bothered to continue watching for whales from the observation point, since the hunters knew that when a narwhal is killed, the remaining animals disperse for a while. But the lighting conditions were spectacular for scanning the fjord, so I spent most of my time there watching the seabirds and icebergs, staring into the shallow water where hardly any living things were apparent, and enjoying the warmth of the Arctic sunshine. I could see and identify birds at a great distance. Most were the same species I'd been eyeing the whole week—lots of fulmars close to shore, with glaucous gulls and black legged kittiwakes swirling in among them occasionally, and numerous groups of common murres and black guillemots further out though never mixed together.

After watching the same common birds for hours, it was obvious to me when something new came into view. A gull-sized bird with a big pale chest, a brown back, and pointed wings pumping hard—a long-tailed jaeger—flew straight and fast through the center of the fjord, dipping a wing occasionally to taunt the nearby gulls. Twenty minutes later its close cousin, a parasitic jaeger, dressed all in chocolaty brown, streaked through, followed closely by two more. The birds seemed less interested in going somewhere and more interested in stealing someone else's food, typical behavior for this family of large seabirds. They chased a couple of gulls without really trying, then changed direction and returned to wherever they came from.

Then, I saw the one bird I had been hoping to see during the trip, an ivory gull, the graceful, pure-white bird that spends almost its entire life above the Arctic Circle. It calmly soared by my perch, hovered in the breeze briefly as if to give me a better view, then continued to follow the coastline west. Better adapted for northern living than most terrestrial birds, it has longer claws than other gulls for gripping icy environments and less webbing on its feet to reduce heat loss. Ivory gulls use dark seaweed in their nests to attract the sun's energy as an aid in incubating their eggs. They are one of the most-wanted birds among the obsessed birding community, of which I am a proud member, and yet I saw only that one despite spending my time squarely in the middle of its breeding range. The reason has only recently become evident.

On the gravel plateaus of northwestern Baffin Island, ivory gulls formerly nested in such abundance that the Inuit sometimes mistook the birds for large patches of snow. Only their short black legs and dark, yellow-tipped beaks gave them away. But in recent years, the birds abandoned thirteen historic nesting sites on the island and, in 2006, just one pair was found during an extensive search of appropriate habitat. That's not the only place from which the gulls seem to have disappeared. Since 1980, ivory gull numbers have declined by seventy to eighty percent, and in all of Canada just 842 birds were sighted during the 2006 survey, prompting the Committee on the Status of Endangered Wildlife in Canada to list it as endangered. They are classified as endangered in Norway and rare in Russia, although their population numbers are unknown in either country.

Ivory gulls feed on fish and marine invertebrates while also scavenging tidbits of carrion from marine mammals and whatever else edible they can find. They spend the winter at the southern end of the pack ice and prefer to nest on inland cliffs and plateaus emerging from glaciers to escape predation by egg-eating Arctic foxes. Even tiny lemmings have been known to eat the birds' eggs. Remote mountain peaks, called *nunataks*, in southern Ellesmere Island and Devon Island, north of Baffin Island, have been strongholds of ivory gull populations in the past. But between 2002 and 2006, twenty-six of Canada's sixty-five historic nesting colonies were abandoned, and most of the remaining colonies declined in size from an average of sixty-eight birds in the 1980s to just nine in 2002. Some biologists predict that breeding populations of the gulls will disappear at most sites within a decade. The one colony that appears to have the best chance of hanging on for the long term is on Seymour Island to the west, where breeding ivory gull numbers have declined by "only" 2.7 percent per year since the 1970s.

What is causing the decline is uncertain, although some speculate that rising Arctic temperatures and changes to sea-ice conditions may be partly to blame. Pollutants may also be a contributing factor, along with competition with larger gull species. In addition, the population declines on Baffin Island coincide with expanding efforts to explore for minerals.

Research scientist Birgit Braune of Environment Canada, who studies toxic chemicals in wildlife, told me via email that no one is certain what is causing the ivory gull population decline. She has tested for chemical pollutants in their eggs and, while concentrations of persistent organic pollutants like PCBs, DDT, dioxins, furans, and industrial flame retardants did not appear in sufficient levels to cause concern, the levels of mercury found in the eggs were among the highest found in seabird eggs anywhere in the circumpolar Arctic. Like narwhals, ivory gulls feed near the top of the food chain, making them highly vulnerable to mercury contamination and likely leading to impaired reproductive success.

Mark Mallory, a seabird biologist with the Canadian Wildlife Service, speculates that the birds are facing multiple stressors that, together, are hitting the birds hard. In addition to contaminants and ecosystem alterations as a result of climate change, he said that ivory gulls were shot in large numbers at least through the 1980s as they migrated along the west coast of Greenland. He notes, however, that while the Canadian ivory gull population has plummeted dramatically in recent years, he believes their numbers in Greenland, Svalbard, and Russia have remained somewhat stable.

"Nonetheless," he told me, "with relatively high contaminants and a continuing shrinking of the sea ice, I think we should expect more stress on the population, which may mean range shifts in the foreseeable future from areas where ice was formerly common but now is sparse."

× × ×

When I came down off my perch around lunchtime, Mads Ole offered me my first bite of muktuk. It was a moment I was hoping for and dreading at the same time, given that whale skin and blubber didn't sound particularly appetizing and I had heard stories of many people from the South having difficulty choking it down. Yet it was an honor to be offered muktuk, and I certainly didn't want to offend my gracious host. By the time the offer came, most of the rest of those in camp were already feasting on it.

Picture an eight-inch square, one-inch thick piece of skin and blubber held flat in your hand, scored with a knife into a checkerboard pattern to make it easy to bite off individual morsels. Mads Ole told me to take a bite and follow it immediately with a piece of dried fish for flavor, and without hesitation, I tossed a piece of muktuk in my mouth, chewed it twice, and added a piece of fish. As I expected, the muktuk itself didn't have much flavor, which is why they combine it with dried fish. The challenge, however, was that it was the consistency of beef gristle or cartilage. It was hard to bite through, so it took a long time to chew before I was able to swallow it. Not being a particular fan of dried fish, the combination was hardly enjoyable. After I finished it, smiled and thanked him, Mads Ole offered me a whole checkerboard of my own, which I politely declined in favor of just one more bite-sized morsel instead.

At mid-afternoon, when I was once again scanning the fjord from the highest point of the lookout, Mads Ole called out to me, "Mr. Todd, come here!" Hidden from view on the far side of Gedion's tent, the hunters had prepared another delicacy that I felt I couldn't decline—narwhal steak. As my wife will attest, I'm not at all daring when it comes to trying new foods, and while I knew this trip was going to challenge my palate, I had no inkling that I would be offered anything other than muktuk. When they set before me a plate full of steaks, and all of the hunters and their wives looked on waiting for my reaction, I must have hesitated a little too long. The steaks looked black and raw just as they did when they had been freshly carved from the dead whale. I wasn't sure I could—or even should—eat raw whale meat, though I don't know why I didn't have the same reaction to raw blubber and skin. But Mads Ole quickly came to my rescue, telling me that the steaks had been cooked and grabbing one for himself with his knife.

With my greatest concern laid to rest, but without a knife of my own, I grabbed a steak with my hands and quickly took a bite. It was too hot to handle, so I quickly dropped it back on the plate and licked my burning fingers as everyone laughed. I didn't realize that the steaks had just come out of the frying pan moments before. When I got past the temperature, I noticed that it tasted very much like liver, a food I hated as a child and hadn't eaten in thirty years. But the narwhal steaks were cooked in garlic butter—which was sold in a green tube in the Qaanaaq market—and despite being the color and consistency of cooked liver, the taste wasn't bad at all. I took another bite. The nervous moment over, Mads Ole wandered away and I stayed to enjoy my narwhal steak with the rest of the group, none of whom spoke a word of English. I listened in on their conversations, smiled when it seemed the right thing to do, made occasional yummy noises, and offered my hearty thanks to the generous hunters before returning to the lookout.

×   ×   ×

Like at the observation point at Bruce Head in Koluktoo Bay, the view from the hunting camp encompassed wide swaths of rock, water, icebergs, glaciers, and melting sea ice. All that ice got me wondering just how scientists identify and describe different types of ice, and whether the Inuit have just as many words to describe ice as they purportedly do to describe snow. When Gretel Ehrlich was writing about Greenland, she found two dozen different words for ice in the native language, including *kaniq, qirihuq, qirititat, nilak, nilaktaqtuq, hiku, hikuaq, ilu, hikuiqihuq, hikurhuit* and many more describing rime frost, freshwater ice, thin ice, new ice, ice on the inside of the tent, melting ice, and so on. The narwhal researchers I met mostly talked about sea ice (frozen sea water, as opposed to icebergs, which are chunks of fresh water broken off of land-based glaciers), fast ice (sea ice that is attached to land rather than adrift), and pack ice (drifting ice that is driven together into a large mass). But scientists who study snow and ice have very different definitions of ice.

Walt Meier at the National Snow and Ice Data Center in Colorado, where I had previously received a lesson in Arctic climatology from his colleague, Mark Serreze, told me that the environmental conditions on the sea have a dramatic effect on the way that ice develops. Unlike freshwater, which starts freezing right at the surface layer, sea ice becomes more dense before it reaches its freezing point, due to the salinity in the water. The coldest seawater "sinks before it can really freeze," explained Meier, "and then it gets replaced by water underneath, which is warmer, of course, because it hasn't been exposed to [the cold air at] the surface. So you get this kind of overturning over several meters near the surface, or even over tens of meters sometimes. Before you can grow ice you've got to cool that whole column or that whole layer down. So it takes more energy to form sea ice than it does lake ice, and it takes longer."

The first stage in the formation of sea ice is called frazil ice, which usually occurs in turbulent weather when needle-like crystals form just below the surface of the water. When these frazils start to merge together, they float to the surface and look somewhat like a slushie drink. These glom together and eventually form a consolidated sheet, and if there's even a little bit of waves it tends to form what scientists call pancake ice. "What happens is that these consolidated sheets start bumping into each other and the edges get ridged up a little bit and tend to be circular in shape," Meier said. "So you get these little circles of ice all over the place that look like pancakes. Eventually the spaces between the pancakes start to dampen any kind of waves and eventually they'll stitch together to create a consolidated ice pack."

In calm conditions, on the other hand, ice forms in a very thin, flat film called grease ice, which looks like an oil slick on the surface. In its earliest stages, when it's still transparent, it's called *nilas*, but as it grows thicker it turns gray and then white.

Meier also said that there is a distinct difference between what is called "first-year ice" and ice that has survived at least one summer season and become "multi-year ice." One difference is in the amount of salinity in the frozen seawater. Newly formed thin ice has the highest salinity levels because it contains tiny pockets of briny water, but those pockets eventually begin to drain out of the ice. During the summer season, the melting that occurs at the surface of the ice flushes out most of the remaining brine, leaving behind ice that contains very little salt.

According to Meier, salinity affects the strength of the ice. "That first-year ice is more flexible, in a way, and in some ways it's even stronger," he said. "Multi-year ice tends to be thicker, but it also tends to be more brittle, whereas first-year ice tends to be more flexible and softer." The age of the ice affects the properties that scientists can see with a satellite sensor. They can often distinguish first-year ice from multi-year ice because the salinity affects the signal transmitted from the sensor.

Given the trends about the warming climate, I asked Meier about the reverse process as well. How does sea ice melt? It sounds like a silly question, now that I think about it. The sun comes out, the temperature rises above freezing, and what was once ice becomes water again, right? Not so fast, he told me. It's true that the process generally begins when the sun comes out—what scientists refer to as the onset of the melt—but the temperature does not have to rise above the freezing point for melting to start.

"That's just a surface thing, just solar heating. We get melt before we see the surface and air temperatures go above freezing," Meier said. "And then you get the melt ponds that form." If you're at the ice edge, he said, the ocean will heat up and the ice will experience a lateral melt that melts inward from the edge. And because currents push the ocean's heat beneath the ice, melting occurs from the bottom of the ice as well, which Meier said has happened at a greater rate in recent years. Typically, he added, sea ice melts at a rate of about one or one-and-a-half meters at the surface for every half meter it melts from the bottom.

These insights about ice would prove useful when my journey to understand the challenges facing narwhals moved to the east.

TEN

TO THE EAST

NOT EVERYONE CAN WITHSTAND THE RIGORS AND EXPENSE OF taking a trip to the Arctic to see narwhals, but for many it may not be particularly inconvenient to see the next best thing. To demonstrate, I took a brief adventure to historic Cambridge, Massachusetts, to a place tucked in among the many stately academic buildings of Harvard University and just a short walk from Martin Nweeia's dental laboratory. It's an impressive brick structure that, more than any other building on campus, is likely to be filled with screaming, running, happy, and enthusiastic elementary school children: the Harvard Museum of Natural History. I arrived early on a Sunday morning, the first visitor to walk through the door, hoping to avoid the crowds and enjoy the quiet while imagining myself surrounded by live versions of the ancient and contemporary creatures housed at the museum. But it was not to be. Moments after my arrival, dads with strollers and moms with their arms full of snacks and sippy cups joined me and a busload of children in exploring the place. Up three flights of stairs, through exhibits of fossils and dinosaurs and fish and insects, I finally found the room I was looking for.

Entering the Hall of Mammals, it is easy to miss the narwhal skeleton, which is located immediately overhead as you walk through the doorway. That's because the room is a festival of fur and bone, and somewhat overwhelming: giraffe and okapi specimens to the left; mountain goats and impala straight ahead; and yaks, gaur, and bison to the right; both skeletal and stuffed specimens of each. A further glance around the hall finds zebras and rhinos, aardvark and anteater, sloths, camels, warthogs, kangaroos, and gorillas. Dominating the space overhead are the complete skeletons of a sperm whale, fin whale, and right whale, the latter still leaking oil from its jaw bone, staining it orange and making it glisten in the artificial light. The baleen of the right whale looked hairy and weathered—I learned later that it's fake—while the long narrow jaw and giant, dinosaur-like teeth of the sperm whale appear oddly out of proportion.

I don't know why I'm always attracted to natural history museums. After all, most of them seem to have the same variety of animals in the same poses with little habitat to place them in context. The museums are often like a police line-up of unusual creatures, all staring into space hoping to be picked to go on a field trip. And live animals, even those just seen on video, are more interesting to watch than the unmoving specimens behind glass at a typical museum, because at least on video one can observe the animals' behaviors, movements, and interactions with their environment. Yet everywhere I travel, the local natural history museum is always near the top of my list of places to visit. This one, because it had a narwhal skeleton, was especially appealing.

When I finally turned around and glanced up at the narwhal, it wasn't at all what I expected. Measuring twenty-three-feet long including a tusk of nearly seven feet, its vertebral column was longer and larger than I would have guessed it would be, and its pectoral fins—looking more like hands than flippers—appeared disproportionally small. From below, its skull clearly showed how its left tooth emerged into a tusk, while on the right there was no evidence that an impacted tooth still lay beneath the surface, making the whole skull look somewhat lopsided. Its ribcage was completely enclosed, with most of its ribs meeting at a central breast bone, similar to the other toothed whales on display but unlike the baleen whales. Looking down the length of the skeleton from the tail end, its body cavity inside the ribs looked small, though the flesh of a live animal and its outer layers of blubber would make up for what appeared to be an unusually skinny animal.

Later I discovered that the male narwhal skeleton on display was probably collected in the 1850s by Dr. Elisha Kent Kane of Philadelphia, a medical officer in the United States Navy and a leading Arctic explorer of the time. Although he died in 1857 at the early age of thirty-seven, his life was filled with adventures on many continents. Described by one historian as "one of the last of the race of brilliant and versatile amateurs," he was perhaps best known for his expeditions to the Arctic in search of John Franklin, a fellow Arctic explorer who disappeared with his crew of 129 men during a search for the Northwest Passage in the 1840s. Kane was the chief medical officer on the first voyage he embarked upon in 1850, and he later commanded a second expedition that took him up the west coast of Greenland farther north than any other explorer had gone before. In a book about his adventures, Kane reported that he gave a "noble specimen" of a narwhal to the Academy of Natural Sciences of Philadelphia in 1851. There are no records of whether this animal was later sold or traded to the Harvard museum or whether Kane donated a different narwhal to Harvard, which also has a beluga skeleton reportedly collected by Kane. Regardless of how it found its way to the museum I visited, the narwhal skeleton is a rare specimen that provides the public with a unique opportunity to envisage this legendary creature.

×　×　×

The range maps printed in marine mammal field guides indicate that narwhals can be found many thousands of miles to the east of where Kane explored and likely captured the museum's narwhal specimen, to regions of the Arctic that are seldom visited and little studied. As suggested by the question marks scattered about some of those maps, that means that little is known about the narwhals living on the east side of Greenland and beyond to Norwegian and Russian waters. Some Danish scientists volunteered to share with me a few insights about those populations and the research they are conducting in their search for answers. Although many of my questions were met with hesitation and variations on "No one knows," the scientists are slowly developing a picture of these genetically distinct narwhals.

Danish researcher Mads Peter Heide-Jørgensen has collected narwhal tusks from archaeological excavations that he believes are 10,000 years old. He is conducting genetic analyses of the tusks and comparing them to present-day narwhals from both East and West Greenland to see what they can tell him about how and when the two populations became isolated. He is also conducting studies to determine the ages of narwhals and measuring how fast they grow, including an effort to see if he can calibrate the ages of the animals by measuring the radioactive elements from nuclear bomb testing that found their way into the layers of their tusks back in the 1950s.

Heide-Jørgensen, who has worked closely with Kristin Laidre for more than a decade, said that he has a mandate from the Danish and Greenland governments to conduct narwhal research in East Greenland. In a telephone conversation from his office in Copenhagen, he said that a 2008 aerial survey of narwhals concluded that about 6,500 animals summer in the southern half of the East Greenland coastline, where he believes the bulk of the population is found. It was the first time abundance estimates have been calculated for the region. Narwhals north of Scoresby Sound, the world's largest fjord system, which is centrally located along the coast, are protected by the Northeast Greenland National Park, the largest national park in the world, which covers 375,000 square miles. Since virtually no one lives in the area and no hunting takes place there, no attempt has been made to count the narwhals that spend the warmer months in the Arctic waters along the park's boundary.

Based on the survey results, it appears that narwhals congregate in summer along most of the south-central coast of East Greenland, with the largest numbers found in Nordvestfjord and Fønfjord in the Scoresby Sound system, as well as in the Kangerlussuaq, Tasiilaq, Sermilik, and Kangertigtivatsiaq fjords farther south. Heide-Jørgensen also found scattered small groupings of narwhals in twenty-two of the thirty-seven fjords along the Blossville Coast between Scoresby Sound and Kangerlussuaq, and they have been recorded as far south as Umivik, but densities are believed to be quite low in these areas. Narwhals are also found east across Fram Strait around the Norwegian archipelago of Svalbard and farther east around Franz Josef Land in the Russian Arctic, but even less is known about those animals, and no one has made any effort to study them.

Whether the East Greenland narwhals are all part of one subpopulation that intermingles year in and year out is uncertain, but it's a question Heide-Jørgensen hopes to answer so he can provide better advice to the government about how much hunting the narwhals can withstand. Just two communities in East Greenland hunt narwhals—Tasiilaq and Ittoqqortoormiit—where an annual hunting quota of eighty-five whales went into effect in 2009, reducing the annual harvest from an average of 110 whales killed during the previous five years. To determine whether all narwhals in East Greenland are part of the same subpopulation, Heide-Jørgensen must gather information on movement patterns and seasonal shifts in occurrence throughout the region.

The first step in that assessment took place in August 2010 when the Danish researcher attached satellite tags to seven narwhals in Scoresby Sound. As Laidre learned in West Greenland, it wasn't easy to tag the animals. "It took us three years to get there, but it was just a matter of getting the right locality," Heide-Jørgensen said. "We did some reconnaissance one year, we decided on the locality the second year, but that locality didn't work out so we tried a different locality in the third year."

The objective, he said, was simply to learn their migration routes, determine their wintering grounds, and see if those locations might bring them into conflict with the most productive commercial fishing grounds in the Greenland Sea. It was the first time anyone had tried to track narwhal movements in East Greenland, and Heide-Jørgensen had only a very general idea of where they would turn up.

"We know they have to leave the coast and go somewhere offshore, but where exactly they are wintering is a complete unknown," he said. "It's a huge area. They could go north to [the Norwegian volcanic island of] Jan Mayen. They could go south. We know approximately where they will be within a 1,000 kilometer box, but we don't know anything more concrete than that."

The tagged narwhals left Scoresby Sound around the first of November and slowly made their way southward to an area of the Greenland Sea between Greenland and Iceland in January. "We don't know if they stay there all winter. What we do know is that they can't get back to the fjords before July because there has to be open water, and that's when the ice breaks up."

The second objective of Heide-Jørgensen's narwhal tagging project had little to do with learning more about the natural history of narwhals. Instead, he is following in Laidre's footsteps by using the satellite tags to collect data on sea temperatures in the Greenland Sea in an effort to provide insight about a globally important climatological process. He said that it is more important to collect sea temperature data in East Greenland than in West Greenland because of the vital role the region plays in what oceanographers call the global oceanic conveyor belt.

Warm water from the Gulf Stream is transported north at the surface of the Atlantic Ocean, where it meets up with cold winds from northern Canada. The layer of warm water then sinks and trades places with a mass of cold water from below, whereupon the warm water turns southward along the floor of the North Atlantic and eventually resurfaces as the world's oceans circulate from one to another. This oceanic conveyor belt is a major reason why northern Europe maintains a moderate climate. If this conveyor were to shut down, as happened for several centuries beginning about 1400, the climate in northern Europe would become considerably cooler. "The global circulation of water masses depends on this heat pump in the Greenland Sea," said Heide-Jørgensen, highlighting the considerable importance of his collection of sea temperature data for use in climate models.

While this aspect of the tagging project isn't directly relevant to his narwhal research, the environment in East Greenland is just different enough from that of West Greenland and Arctic Canada that it's certainly worthwhile to pay it extra attention. The ice conditions are reported to be "rougher" in the East, and the ice season lasts longer there as well. It's also a less productive ecosystem with a far less diverse marine environment compared to that in West Greenland. Heide-Jørgensen calls the waters of East Greenland "a desert, with fewer fish species available to feed on," adding that this may not matter much to the narwhals that live there, however, since they feed upon a limited variety of prey items anyway.

In addition to these challenging environmental conditions, narwhals in East Greenland may face even greater threats to the health of their populations than do those in West Greenland and Canada (with the exception of hunting, which is perhaps less of a threat in East Greenland since only two communities hunt narwhals there). The threats caused by pollutants, especially mercury, are especially elevated in East Greenland due to the atmospheric conditions that drive industrial emissions from China and elsewhere in the Far East to the region before they make their way farther to the west. Environmental concerns over oil drilling are greater in East Greenland, too, since it is believed that the seafloor off the east side of the island holds twice as much oil as in the West. There are likely to be far more oil drilling operations in the East in coming years, which will bring with them significant threats to narwhals from the inevitable spillage of oil and other pollutants, as well as disturbances caused by noise from the industrial activity. It's all quite worrisome to the narwhal biologists, but there is little they can do now except gain as much knowledge as possible about the status of the population so comparisons can be made about their health in the future.

× × ×

Biologist Rune Dietz of Denmark's National Environmental Research Institute is studying one of the most worrisome issues: contaminants in the marine environment that find their way into the narwhal's vital organs, blubber, and other tissues—even their tusks—via the food chain. While no studies have yet been conducted that have evaluated the health effects of contaminants like heavy metals and industrial chemicals on narwhals or other Arctic marine mammals, several studies are under way to quantify the contaminants and compare them in various populations of the animals. In a study of tissue samples collected from 150 narwhals from several different locations in Greenland in the 1990s and 2000s, Dietz found elevated levels of cadmium, selenium, and mercury, as well as man-made organochlorines like PCBs and DDT. Female narwhals were found to have significantly higher concentrations of heavy metals than males, and he noted that concentrations increased during the first three to four years of their lives before leveling off. Organochlorine concentrations also varied by age and sex, with females showing decreasing concentrations in the first eight to ten years of life, while males saw increases in their first few years.

Dietz said that the results were similar to studies conducted of narwhals in Canada, which would be expected since West Greenland animals mix with Canadian narwhals on their feeding grounds in Baffin Bay and so are likely to consume similar contaminants. What I found noteworthy, however, was that PCB and DDT concentrations in West Greenland narwhals were only half that of narwhals in East Greenland, though Dietz had a reasonable explanation for why. "If you look at water transport in the Arctic, the water masses actually come from the east coast of Greenland and go toward the west. It could be a kind of dilution [of contaminants] that takes place as the water moves west," he said. His results are consistent with studies of polar bears and ringed seals, which also contain higher organic contaminants in East Greenland animals compared to those in West Greenland.

All of these contaminants travel on prevailing winds in the atmosphere and are deposited in the Arctic in the rain and snow. PCBs and DDT are volatile compounds that evaporate at lower latitudes and are transported to the colder regions, while many of the heavy metals start out as emissions from coal burning power plants. While concentrations of many of these contaminants in the environment are slowly decreasing over time, mercury—perhaps the most dangerous one because of the known deleterious effects it has on the brain and reproductive system—is increasing.

"It seems like the hotspot for the emission of mercury is coal burning in China, which has seen a dramatic increase," Dietz said. "And that fits well with the increases we've been seeing in the Arctic. It is transported through the air masses, and on an elemental level mercury can stay up there for several years before it is precipitated into the Arctic environment." About 200 tons of mercury finds its way into the Arctic region each year, about 10 percent of the world's emissions of mercury. And the decline of multi-year sea ice due to global warming means that mercury that may have settled on the ice and was having little effect on wildlife is now finding its way into the marine environment where it is accumulating in the tissues of whales and other creatures.

Dietz is in the midst of several additional studies of mercury contamination in Arctic marine mammals. He has found that polar bears generally have low levels of mercury in their brains because they can rid their system of a substantial amount of mercury through their fur, a process whales cannot duplicate since they have no fur. When we spoke, Dietz was preparing to dissect some narwhal brains to assess their mercury levels.

He is also analyzing mercury in the growth rings of narwhal tusks, which, like hair, teeth and feathers, can be a storehouse of pollutants. "I can go out and get a two-meter tusk and get the last fifty-year history of that animal's mercury contamination," Dietz said. The challenge, he noted, was distinguishing whether the year-to-year mercury changes found in the layers of the tusk represent changes in the amount of mercury in the environment or whether they are due to a life-long bioaccumulation of contaminants. An analysis of individual narwhals of different ages should provide insight into that question.

Dietz isn't the only scientist examining narwhal tissue samples to better understand the health of the population, however. With the support of the Nunavut Wildlife Management Board, University of Manitoba scientist Gary Stern has used Inuit hunters to collect samples of narwhal liver, kidney, muscle, and blubber to assess contaminants in whales in the Canadian part of their range, and his results somewhat mirror those of Dietz. He said that climate change may be exacerbating the problem because the accumulation of contaminants in narwhal tissues is dependent on the whales getting access to those contaminants. As sea ice retreats, he said, more contaminated fish will be available for the whales to feed upon, making those contaminants "bioavailable."

Stern said that contaminant levels in narwhals are similar to those of belugas in the eastern Arctic, which also eat fish high in the food chain. (Belugas in the St. Lawrence River population have contaminant levels 100 times the rate of eastern Arctic animals.) Bowheads and walruses, which eat zooplankton and mollusks, respectively, have lower contaminant levels because their food is at a lower trophic level. While Stern agrees with Dietz that little is known about the health impacts of these pollutants, he worries most about what he calls the "synergistic effects" from a wide range of challenges the animals are facing.

"It's hard to tell what effects the contaminants are having on their health, but they are one additional stressor they have to deal with," he said. "We still have no information that says directly that it's affecting reproduction or having neurological effects; it's hard to tell with an animal in the wild. But these animals are under stress for a number of reasons—changing habitat, noise pollution—and contaminants are just another thing that acts synergistically to possibly make their immune systems not work as well."

While contaminant levels in the environment are lower than in recent decades because, he said, "we've shut off the tap and fewer contaminants are traveling north," Stern is not expecting much of a decline in contaminant levels in marine mammals in the near future because the chemicals linger in their tissues for many years.

×   ×   ×

Although it is much more difficult to quantify, the threats resulting from the effects of climate change, including ecosystem shifts, ice loss, and increased industrial activity, may be just as worrisome for narwhal populations as the build-up of contaminants in their bodies. And the latter factor appears to be the first it will face.

According to the *Wall Street Journal*, the U.S. Geological Survey believes that there may be as many as fifty billion barrels of oil off the coast of Greenland, a quantity similar to that of Libya, which could provide an economic boost that would transform the island and sever its financial dependence on Denmark. The expense of drilling in the region, along with the short drilling season and the risks that oil rigs would get rammed by icebergs, had made oil drilling in the region unattractive. But increasing oil prices, satellite tracking of icebergs, and warming temperatures have created a huge opportunity and growing interest among oil companies around the world. The political wrangling over what country controls the rights to oil and gas and other minerals at the North Pole is a testament to the great potential wealth to be found there. In 2010, a British company drilled the first exploratory oil wells off Greenland in more than a decade, and the same year the government awarded more than a dozen new exploration licenses to companies from several nations.

These developments are alarming to a number of groups, including those concerned about the health of narwhal populations. Narwhals are skittish, and the noise and increased shipping resulting from growing industrial activity is certain to be a significant disturbance to their feeding, migration, and communication, to say nothing of the threat of oil spills, which could be devastating. The oil industry admits that their ability to respond to an oil spill in Arctic waters will be hindered dramatically by the small human populations in the area and the limited number of boats that could be available to work skimming the ocean surface. During the Deepwater Horizon oil spill in the Gulf of Mexico in 2010, about 30,000 volunteers—equivalent to the entire population of Greenland—helped to clean beaches, remove contaminated seaweed, and rescue oil-covered wildlife, and more than 4,000 local boats were hired to skim oil from the ocean surface and undertake other recovery and response measures. Nowhere near that many people and boats would be available to help in a spill off Greenland or other areas of the High Arctic, leaving the ecosystem and its wildlife to suffer the consequences. It disturbs me to read about the gold rush the oil companies are undertaking to drill for oil in the Arctic with too little thought and resources devoted to the inevitable damage they will cause from what will likely be innumerable spills, despite the outcry that resulted from the inept response to the spill in the Gulf of Mexico, where clean-up conditions and manpower were nearly ideal.

Representatives of the fishing industry in Greenland have also publicly expressed their concerns about the effect that oil exploration will have on commercial fishing, though some narwhal biologists worry about the fishermen, too. It is predicted that the loss of sea ice will open up additional areas for expanded fishing for halibut, the primary prey of narwhals and a species with snow-white meat that is high in fat and rich in flavor, making it popular for sushi, sashimi, and smoking. It's a worrisome possibility, especially considering that when commercial interests compete against wildlife for a resource, the animals almost always lose.

Narwhals have a very restricted diet, with Greenland halibut being their predominant prey. Halibut are widely distributed in the North Atlantic, and commercial fishermen traditionally harvested about 20,000 tons per year in the fjords of northwest Greenland using longlines and gillnets. An additional offshore fishery developed in the 1990s in Davis Strait that now captures more than 10,000 tons of halibut each year, and efforts are underway to open up additional fishing grounds in the deep waters of central Baffin Bay where the narwhals spend the winter feeding. The more the ice recedes, the more fishing is likely to take place, leaving fewer and fewer resources for the narwhals.

On the other hand, some biologists suggest that the decreasing ice pack may open up new areas for narwhals to exploit, including areas far from human settlements where they would be free of hunting pressure. And while increased fishing and oil and mineral extraction will certainly be a disturbance to narwhal populations, it may also provide high paying jobs to Greenlanders who previously had depended on hunting narwhals for subsistence.

The true impact of these factors on the ice whale remains unknown.

ELEVEN

PLAYING CATCH

THE LOGISTICS OF MOUNTING A MARINE MAMMAL RESEARCH project in the Arctic are daunting. When I was invited as the thirteenth member of a research team whose aim was to capture live narwhals, I had no idea what I was getting myself into. I just knew that in my search to understand the narwhal and its melting world, I needed to see the animal through the eyes of a researcher in the field. That's exactly what Jack Orr provided me.

A biologist with the Canada Department of Fisheries and Oceans, Orr has been leading research expeditions to the Arctic for more than twenty-five years, more than forty-five projects in total. A wiry man in his fifties with piercing eyes and thoughtful features, he started out in the 1980s tagging beluga whales by chasing them into shallow water with inflatable Zodiac boats, tossing a large hoop net over their heads, and jumping into the water to wrestle them to shore. As he described the process, I couldn't help but picture the antics of the Keystone Kops attempting to handcuff a struggling whale, only to find that they had bound and gagged themselves on an ice floe. In deeper water with more difficult access, Orr developed a method of surrounding a group of belugas with a long net akin to a heavy-duty fisherman's gillnet. Since the mid-1990s he has been capturing narwhals in nets using a technique that he casually described to me as "just throw it out there and hopefully you catch something." Clearly there is more to it than that. Orr has refined a somewhat simple narwhal capture method he first saw Danish scientists try that involves placing a long net across an area where narwhals are known to travel. While it requires a bit of luck and a great deal of trial and error at each different site, Orr has become known as one of the world's leading experts at the live capture of Arctic marine mammals.

It took several short flights by a Twin Otter aircraft to ferry the research team's gear and personnel from Pond Inlet to a cobble beach in Tremblay Sound, a long fjord about 70 miles to the west. There were also two flights carrying gear from the village of Resolute Bay on Cornwallis Island, where the Canadian Polar Continental Shelf Project coordinates logistics for dozens of Arctic research projects every year. Two Inuit boats ferried additional supplies and personnel to the campsite as well. It wasn't until I was helping to load the gear on the plane that I realized the overwhelming job of coordinating such an endeavor. Among the equipment we crammed into every corner of the plane were two deflated Zodiac boats and three outboard motors, two fifty-gallon drums of diesel fuel, a twenty-five-foot tall radio antenna, three gas grills and propane tanks, a fancy vessel containing liquid nitrogen for freezing tissue samples, a dozen sheets of plywood and twenty two-by-fours, several huge nets, and a wide array of scientific paraphernalia. That doesn't even include the tents, heaters, and personal gear for thirteen people. Then there was the food—1,800 pounds of everything imaginable, including two legs of lamb, dozens of steaks, about as many chickens, a case of wine, eight cases of soda, lots of fresh fruits and vegetables, and even a bag full of chocolate bars. Orr sure knows how to keep a team of researchers happy in cold and uncomfortable conditions. It was all a huge relief to me, as I had worried about being cold and hungry for my two weeks in camp. As much as I love the scenery and the wildlife and the solitude of the Arctic, I get cold easily, which could have made for a miserable trip.

I was on the last flight leaving Pond Inlet, with my knees at my chest, my arms resting across a wheelbarrow, and 5,000 pounds of gear in front of me no more than a foot from my face, worrying the whole time whether we were overtaxing the small plane's engine with the heavy load. Doing my best to clear those concerns from my mind, I found that if I twisted my body just right I could glance out a tiny corner of a tiny window, where the view of Eclipse Sound and the glaciers on Bylot Island was mesmerizing. While it wasn't officially an aerial survey, it was the first time I could stare down from 1,000 feet up to count the whales swimming below. We saw only a few small groups of narwhals on the thirty-minute trip—no bowhead whales, much to my chagrin, since I had been told that the flight would be my best chance to see one—and the animals were much harder to see than I had guessed, giving me a greater appreciation for the job of the real aerial survey teams.

As we neared the end of the flight, I couldn't help but feel giddy about the adventure that was unfolding, though I did my best to hide it and the accompanying nervousness that always reveals itself regardless of how many research camps I visit. The landing was loud and bumpy, and the runway—such as it was—was less than 200 yards long, so I was pleased that my first flight in something other than a commercial jet ended successfully. It wasn't until we had unloaded the gear that I had an opportunity to glance around and take in the place— rocky hillsides jutted up from the water's edge, hints of soft green were scattered amid innumerable shades of brown, glaciers glinted white between the rocky crags, and a few small icebergs floated by like the blessing of a fleet of oddly shaped boats.

"So much of the Arctic has to do with blindness and seeing: we go snowblind when the sun is too bright, as if too much exposure to what is real exhausts us," wrote author Gretel Ehrlich of her travels in the region. "We let ourselves be fooled by Arctic mirages. . . . There is ice blink and water sky; there are shadows miraging shadows, cloud layers that shroud snowfall, and ice through which you can see water. And in the dark months the mind's troubled eye opens wide, a cerebral lens that allows the imagination to blossom." Although I had first read these words several months before, the images they had created in my mind were resurrected in those first hours in camp.

We spent the next day and a half setting up camp—turning an existing wooden shack into a science lab, setting up a Quonset hut to use as a kitchen, inflating Zodiacs, tuning up motors, and prepping the scientific equipment. It was harder work than I expected, and I had few of the handyman skills that would have speeded the project along, but it helped to turn the group of colleagues, acquaintances, and strangers into a cooperative team. I felt like whistling the Seven Dwarfs theme song all day.

While all of this work was going on, we glanced up in unison as our first massive movement of narwhals swam by. At first we just heard gentle splashing noises, like a child in a bath, as they surfaced for air, and an occasional whooshing sound as they exhaled. The whales came into view, swimming in groups of threes and fours and sixes, most having the dark coloring of younger whales, some with pinkish gray calves that looked like tiny swimming sausages, and all porpoising at the same speed in the same direction with little concern for the colorful assemblage of onlookers standing on the shore less than fifty yards away. I watched in stunned amazement, barely able to catch my breath, so pleased was I with their apparent tameness and abundance. There must have been 200 of them, none with a visible tusk, and they took more than ten minutes to pass our camp. If the scientists had worried about whether narwhals were going to be in the area, those worries quickly disappeared. Little did we know then that what we believed were the same animals would swim by us repeatedly two or three times a day for the next few weeks.

When camp was established, the next project was to set out the nets, which was a more challenging task than any of us had imagined. It took six people in both Zodiacs almost an hour just to set the anchor for the buoy that connects to the net seventy-five yards from shore. We learned later that the water was 193 feet deep in that spot, and the current was so strong that it kept dragging the anchor, which consisted of a couple of bags of sand and rocks. Once the anchor seemed to be firmly in place, the net—almost unmanageably bulky and heavy at 100 feet long and fifteen feet high, with weights at the bottom and buoys at the top to intercept any animal swimming near the surface—was unfurled and tied with a long rope to a boulder on shore.

And then the waiting began.

×　×　×

The main objective of the three-week research project was to attach satellite tags on the dorsal ridges of up to ten adult narwhals to track the movements of individual animals. "You can count whales from the air, you can count them from land, you can see what the herd is doing, but you don't really get an idea of what an individual does on a daily basis [without tagging them]," said Orr. The tags tell the scientists where on Earth the whales are located and where in the water column they are spending their time. This data, along with information about water depths and the animals that live at that depth, enables the researchers to create a dive profile, noting how long a whale is at a certain depth and what it may be doing over the course of a year. It gives them insights into not only its movements but also its behavior.

"It's really important to be able to say that these whales in this location only go to these certain places, so they're only hunted by people in these particular communities," Orr added. "It helps our management people to determine the number of animals in certain stocks, which lets us know about hunting pressures. Where the whales are at certain times of year lets us know when they are vulnerable to certain types of human interactions—ice breakers or tankers or fishing or even noises like seismic activity, mining activity on land. . . . You couple that with our data on DNA, analysis of contaminants, and information from hunting, you put all the pieces of the puzzle together, and you understand the animal better."

The tags Orr uses are linked to satellite receivers that are part of the U.S. National Oceanic and Atmospheric Administration's weather system. Not much larger than a disposable cigarette lighter, the tag is attached to the dorsal ridge of the narwhal with two or three nylon pins that are inserted through the whale's skin and blubber to hold it in place.

"With a narwhal being so streamlined, there aren't a lot of places we can put a tag on," he said. "We've tried straps. Some guys have tried attaching things to their tail. [The dorsal ridge is] not ideal, because the whales do use their back to break through the ice, but there really is no other way to put a tag that lasts as long on narwhals and belugas." Scientists have established preferred methodologies for the application of a satellite transmitter on nearly every type of marine mammal. Orr said that he uses epoxy to affix transmitters to the backs of seals and to the back of the head of walruses, for instance.

The tags used on narwhals have a small antenna that communicates with the satellite when the animal is at the surface, and the data collected by the tag is accessible to Orr any time via a secure website. The device is powered by a C-cell battery, which gives it a lifetime of about six to twelve months and up to 125,000 transmissions. Eventually the pins slide out, the tag falls off, and the holes in the dorsal ridge close up like a pierced ear after its earring has been permanently removed.

Orr has tagged narwhals in about a dozen different Arctic locations over the years, and each site presents its own set of challenges, including strong winds and currents, shallow water, distracting noises, and nearby hunters. He has typically tagged four to eight animals each year, though he has been shut out completely on a couple of occasions. Drawing significant conclusions about narwhal movement and behavior patterns requires a great deal of data from a large number of whales, so it takes a long time to get the information he seeks. Tagging data played an important role in the discovery that narwhals typically spend the winter in areas that are 98 percent covered in ice. "That's showing us that these animals are in a very fragile, ice-loving environment, and any changes to that environment could have dramatic effects on them," Orr said. "Little changes in global climate change or ship traffic or even the commercial harvest of turbot (halibut), which they seem to be feeding on in the winter, can have a drastic effect on the population."

One question Orr hoped to answer with the tagging project was whether there is any intermingling between the narwhals that summer in the fjords around Tremblay Sound and those that summer 150 miles away in Admiralty Inlet, the largest inlet in western Baffin Island. Some researchers have hypothesized that the two populations are distinct stocks and should perhaps be managed independently. As Orr said, "even though they're in the same general area, never the twain shall meet, maybe."

×   ×   ×

Once the net was in the water, the researchers were divided into two-person teams to stand watch for three-hour intervals. The job was partly to keep an eye out for polar bears, which weren't expected to be in the area at that time of year, and mostly to note when narwhals were approaching the net and to call out to the rest of the team if a whale became entangled. The hardest part of the job was staying warm and staying focused, since on most shifts there was little activity and it was easy to get caught up in the life of the camp. When narwhals were sighted in the area, the group was put on alert and began donning their gear with the expectation that a whale would be captured within minutes. Everyone had their assignments—some were responsible for jumping into the Zodiacs and racing out to the net to raise the entangled whale to the surface so it could breathe, other team members were assigned to pull the shore-end of the rope to drag the boats, net, and whale to shallow water, while the scientists prepped for their research. My job was as a rope puller and assistant to tusk researcher

Martin Nweeia, but my watch-duty partner was Sandie Black, head veterinarian at the Calgary Zoo, whose primary responsibility was to monitor the health of the captured narwhals and collect blood and tissue samples.

During our first couple of shifts, we saw no whales, and while the masses of narwhals continued to travel by camp every eight or ten hours, they always seemed to barely avoid the net. At those times, with the team half dressed in diver's dry suits, chest waders, and a wide variety of cold weather gear, we waited with great anticipation every time a narwhal seemed destined to hit the net. I was genuinely nervous, continually recounting in my head my role, determined to make a useful contribution to the project and not embarrass myself or make Orr and Nweeia regret inviting me to join them. I was the outsider of the group, and I didn't want to let them down. Yet everyone else seemed calm and ready. Several times I heard Orr mumble, "come a little closer, just a little closer," but still the animals missed the net.

The enormous net was supported by seven equidistant buoys, and during my watch shifts I found myself repeatedly glancing at the buoys, counting them to make sure a whale hadn't become entangled and taken one down, looking away at something else, and then looking back and counting the buoys again. I must have counted those buoys several hundred times during my two weeks in camp, and never was a buoy missing. That was frustrating for me, though I also worried that I would freeze in place if a whale hit the net and I wouldn't know what to say or how to respond. So I practiced in my head yelling "Whale in the net!" as I had heard others say was the appropriate first reaction.

At 9 p.m., just minutes into our third shift, the narwhals began to return. While most appeared to be on the far side of the fjord, nearly a mile away, the rest of the herd seemed to be moving closer to our side, and eventually everyone in camp was watching with anticipation. It looked like this time it was going to happen. When Orr began putting on his gear, so did the rest of us. Just as I pulled up my waders, someone whispered, "He's coming right at the net. He can't miss it." It appeared impossible for the narwhal to avoid becoming entangled. We watched its wake, and five feet from the far third of the net, it stopped and retreated before going around the net. We let out a collective groan. A short time later, it happened again. A whale surfaced about fifteen feet from the net and continued to approach, but when it was within an arm's reach of the net, the animal stopped, backtracked, and dived under, only to appear again on the opposite side. Orr wandered away to make coffee, and the rest of us stripped off our gear in frustration.

The activity continued that way for a week, during which the period of twenty-four-hour daylight at that latitude came to an unannounced end and a few brief moments of actual "night" occurred. The first day that the sun officially sets in the Arctic is hardly noticed, since it drops below the horizon for barely a minute or two before returning, and light levels are still adequate for all activities. In fact, the period when the sun dips as much as six degrees below the horizon is called civil twilight, a legal definition indicating that all "operations requiring daylight" can continue. The next six degrees below the horizon is referred to as nautical twilight, when the stars used for celestial navigation become visible, followed by astronomical twilight just before true darkness occurs. And while the sun set behind the steep slopes of the Tremblay Sound fjord well before its legally designated time, the modestly changing light patterns did little to affect our operations, other than to inspire us to bring out our cameras to record the subtly shifting hues.

The narwhals continued to travel back and forth in front of us regularly, but they always avoided the net. It was clear to us all that it was not accidental. They had to have been able to see or otherwise detect the net and intentionally swam around or under it. Orr and the rest of the team tried a variety of strategies to increase our chances of catching a narwhal—relocating the net to deeper or shallower water, changing its angle in the water, covering the white surface buoys with black plastic, replacing the net with one made of thinner twine, and even cutting holes in the net—but the narwhals continued to avoid it.

Once, instead of simply traveling from one end of the fjord to the other, the narwhal parade stopped in front of us and began what looked like play. We had debated during the preceding days what the whales had been doing or where they were going, because they always seemed to be migrating one way or the other, never stopping to feed or do anything worth noting. This time they had something else on their minds. While we continued to watch the nets, we observed our first bit of tusking behavior by a few narwhals on the opposite side of the fjord. Two tusks were pointed skyward and were apparently rubbing against each other, though they were too distant to tell for sure. Some animals rolled on their sides and waved their flippers in the air, sometimes seeming to splash their pod mates. Then several more groups of two to four whales raised their tusks at various angles. Through a telescope, Nweeia thought he could see that one tusker was a female by its narrower and shorter tusk, tighter spiral, and whiter coloration. Another male raised his head straight up out of the water by two feet or more in a behavior called spy-hopping, as if he was trying to raise his tusk higher than the rest. Then two tusks crossed over a flipper, making it three animals together in a social interaction that must have had two narwhals side-by-side vertically in the water with a third horizontal between them, something that one of our group suggested may have been foreplay. This mass of activity continued for forty-five minutes as different combinations of behaviors got the research team more and more excited, even the veteran scientists. And then, as always seemed to happen, the whales stopped like a switch had been turned, returned to their usual formation, and traveled together toward the mouth of the fjord. When they were gone, the water became calm, a silence returned to camp, and the researchers went back to their chores.

×   ×   ×

The research team for the project was quite varied in personality, backgrounds, and experience, which kept everyone entertained while we waited to catch a narwhal. There were pranksters and storytellers and shy ones and loud ones; there were the mechanically inclined, the good cooks, the fishermen, and those, like me, whose primary useful skill was washing dishes. The team included four Inuit from Pond Inlet who served as whale handlers, boat wranglers, and firearms experts (in case of polar bears), and they provided considerable knowledge about local geography, wildlife, and weather and water conditions. It was their first participation in a narwhal research camp, which was also true for graduate student Cortney Watt and recent graduate Joe Guay.

Tusk researcher Martin Nweeia, perhaps the quietest member of the team but whose quick wit occasionally surfaced, returned to continue his work to assess the sensory capabilities of the narwhal's left tooth. He again planned to introduce a high salt solution to the tusks of captured narwhals and use a Holter monitor, a device used by cardiologists, to measure changes in heart rate. He abandoned his efforts to also monitor brain wave activity, figuring that it was more important to get consistent confirming results of heart rate changes. "We really need something that's going to hit you between the eyes and say, 'Look, this is the stimulus, this is the response, black and white.' If we get to that level, we've taken a significant step in doing what we wanted to do, which is show that this is a sensory organ."

Veterinarian Sandie Black, my partner on watch duty and one of only two women in camp, has also been a regular participant in Orr's research projects, having joined him in the Arctic in six of the previous seven years. It had been her idea that a vet should be part of the research team "to bring an aspect of health assessment to the work." She was there to monitor the health and stress of the narwhals during the tagging process and to conduct physical exams of the animals at the same time. She said that, for the most part, the animals are quite calm when they get to the shore and struggle very little.

"The ones that struggle the most tend to be females that we assume to be with calves or their pods are waiting for them, but by no means is it constant struggling," she said. During the short period of time when the animals are captive, Black said that the acidity and carbon dioxide levels in their blood drop, and their oxygen levels rise, all of which suggest that the narwhals are recovering from the stress of being captured. "Even though they are still being handled, they are recovering from that initial struggle, and they are in better shape when we let them go than when we first brought them to the surface after capture," she said.

Besides monitoring the narwhal's health during the tagging procedure, Black also collects blood and tissue samples for later analysis. "We don't really have any information about the normal physiological blood values of a narwhal, whether cellular values, red cell indices, white cell counts, blood chemistry. There are no narwhals in captivity, there are not likely to be, and those basic physiological measures give us a starting point for assessing health in the population." She also collects samples that can be used later to assess environmental contaminants, disease exposure levels, and chronic stress.

Clint Wright, tall and jovial with an optimistic attitude that helped raise spirits as the days dragged on, rejoined Orr's research team after a lengthy hiatus and was excited to get his first hands-on experience with narwhals. A former dolphin and beluga whale trainer who helped Orr on his early beluga capture efforts, he is now senior vice president at the Vancouver Aquarium, the only modern facility that has attempted to display live narwhals. That was back in 1970, when the aquarium's founding director Murray Newman organized an expedition to Koluktoo Bay to capture up to a half dozen narwhals to begin a captive breeding program. After weeks of effort, the team caught two adult female narwhals and three calves using a net system similar to that used by Jack Orr. The aquarium also purchased a young male caught by Inuit hunters at Grise Fjord, a native community on Ellesmere Island to the north. (A year before, Grise Fjord hunters captured a baby narwhal, and staff at the New York Aquarium had it sent to their Coney Island facility, where it died a few months later.) Packed in ice and constantly sprayed with water, the animals were transported at great expense and effort on chartered planes back to Vancouver, where they were met by crowds of well-wishers and made front page news throughout the region.

The excitement was short lived. The three calves soon died of pneumonia. "We suspected the warmer water in Vancouver was an underlying cause of death, although our arctic belugas were doing fine under the same conditions, but cooler water was not obtainable," wrote Newman in his memoir, *Life in a Fishbowl: Confessions of an Aquarium Director*. "When it became obvious the babies were sick, an oxygen tent, antibiotics, and steroids were administered in an attempt to save their lives." Two months later, the smaller of the adult females died of a bacterial infection, and the other died of the same cause a few weeks later. The last narwhal, the male bought from the Inuit in Grise Fjord, died on December 26, 1970, just four months after being captured. It was nine feet, seven inches long, weighed 900 pounds, and had been eating fifty pounds of fish each day and appeared healthy. But an autopsy found that he had an abscessed lung.

That disaster put a temporary end to the aquarium's interest in displaying narwhals, though in 1989 they considered helping the local residents of the Baffin Island community of Arctic Bay establish a holding tank for narwhals that could later be transported to aquariums around the world. That project never came to fruition, though it hasn't stopped other aquariums from quietly considering the exhibition of live narwhals in their facilities.

"They did the best they could at the time, but we would do things differently now," Wright told me, noting that the Vancouver Aquarium has agreed not to capture any more wild cetaceans for display. The facility will now only acquire animals that were born in an aquarium, animals that were captured before 1996 (when its non-capture policy went into effect), and animals that have been rescued from unhealthy circumstances.

I sometimes struggle with the idea of captive animals of any sort, whether in aquariums or zoos or other facilities, despite the fact that I enjoy those venues immensely and know the important role they play in breeding rare species and educating the public about wildlife conservation. When I asked whether he thought it would be appropriate for any aquarium to display narwhals, Wright said that aquariums "want to display animals that you can take good care of. We believe the animals [in the Vancouver Aquarium] are well cared for and adapt well. Nobody knows whether narwhals could adapt well to an aquarium situation. An animal that has a tusk would be a unique challenge. When a beluga swims, they swim slowly, they like shallows, they turn around, they're at home in an aquarium setting. But narwhals tend to swim in straight lines, they dive to depth. Are they suitable for an aquarium setting? I don't know."

Wright did say, however, that there would be one clear benefit from exhibiting narwhals. "We've learned a great deal about belugas and killer whales and dolphins in aquariums, and there is still so much we don't know about narwhals from a pure science standpoint. The chance to get close to these animals to learn the little bit we can is incredible. And you never know when that information is going to come in handy, especially as we see with environmental change."

×   ×   ×

While beluga whales appear to thrive in the confines of dozens of aquariums around the world, their status in the wild varies dramatically depending on their location. The closest relative of the narwhal, the beluga or white whale is considered somewhat common over much of its range, especially the populations in the Beaufort Sea, Bering Sea, and Hudson Bay. Beluga numbers in the St. Lawrence River in Quebec and in Cook Inlet, near Anchorage, Alaska, however, are dwindling quickly and may not be able to recover, largely due to pollutants in their environment and the effects of human activities. Unlike narwhals, each of the twenty-five subpopulations of belugas scattered throughout the circumpolar Arctic and sub-Arctic region is genetically distinct.

While belugas and narwhals are the only members of the *Monodontidae* family of small toothed whales and are approximately the same size, they are nonetheless strikingly different. Belugas are slow swimmers and shallow divers that prefer inshore waters where their reputation for being particularly "talkative" led to their common nickname, the sea canary. As adults they have completely white skin, "smiling" lips, a full set of teeth, a fatty and oily melon on their foreheads, and they can turn their head, making belugas anatomically unique and popular with aquarium visitors, including me.

I went to Mystic Aquarium in Mystic, Connecticut, in the spring of 2010 to have a close-up encounter with beluga whales, and it was everything I had hoped it would be. I had been to the aquarium several times previously, and Renay and I had vacationed in Quebec one summer with the primary objective of seeing the wild belugas that congregate in the St. Lawrence River near the community of Tadoussac, but my return visit to Mystic was a truly hands-on experience.

At the aquarium's Arctic Coast exhibit—a one-acre outdoor habitat consisting of three interconnected pools containing 750,000 gallons of water—three belugas were cavorting in the 50°F water as a constant flow of visitors watched from viewing areas above and below the water line. I stared intently through the glass from the underwater observation deck as one seven-year-old whale carried a deflated orange beach ball in its mouth and on its melon, seeming to enjoy the challenge of balancing it as the animal surfaced for air and dived to the bottom. A second beluga circled the tank repeatedly, swimming sideways on the curves and turning its head to look at the tiny humans who were giggling with delight, while a third beluga kept mostly to itself out of sight. As I watched, I was struck by two elements of their physical stature that were somewhat unexpected: their flippers appeared tiny compared to their body size, and their tail stock narrowed considerably before sprouting an unusually small tail with what looked like gray leaf-veins clearly visible. Their dorsal ridge, which they use to break through thin layers of ice, was also much more prominent than I expected, giving it a distinct hump that seems more obvious than the one found on humpback whales.

By then it was time for me to join Justin Richard, the aquarium's beluga trainer, who fitted me for chest waders and took me to the rear of the beluga tank. To study the captive animals, Richard trains them to exhibit behaviors that make it easier to monitor their health, which also makes it safe for small numbers of aquarium visitors like me to wade into the tank with the whales.

As five assistants went to feed the other belugas, Richard signaled to Naku, a twenty-nine-year-old female beluga, to join us in a protected pool. Weighing about 1,350 pounds and measuring eleven feet long, she looked huge to me at arm's length, but her engaging "smile" and precise responses to Richard's cues quickly put me at ease. Richard proceeded to tell me about various aspects of Naku's physiology, starting with her teeth, which are perfectly cylindrical with flat tops and are spaced so the top teeth fit in between the bottom teeth when her jaw is closed. And then he told me to put my hand in her mouth and touch her teeth. I knew I was going to get to touch the animal, but I figured I was only going to be allowed to rub its skin, so I was unprepared to reach in Naku's mouth just moments after entering the water. After a slight hesitation, I tentatively extended my arm to touch her teeth and gums, which felt like mine do, and then patted her tongue, which reminded me of a slimy chicken breast fresh from the refrigerator. If I were honest, that's probably what mine feels like, too, only warmer. With a twitch of his hand, Richard signaled Naku to curl her tongue and raise and lower it to demonstrate how belugas are able to easily reposition fish in their mouths for easier swallowing.

Beluga whales often find food in the seafloor by shooting a jet of water from their mouths and then using a powerful suction to inhale the uncovered food. Richard asked me to form a circle with my hand and place it underwater a foot from Naku's mouth, and she demonstrated her remarkable vacuum ability by sucking my hand right into her mouth. It was so powerful that my automatic reaction was to jump backwards, but Richard said that the aquarium belugas have been trained to demonstrate their suction in a more gentle way than they use in the wild so they don't scare the visitors. If what I experienced was gentle, then her full-force suction would likely have consumed my entire arm! Belugas swallow their food whole, but they use their interlocking teeth to grasp their prey or to break apart hard shelled invertebrates before consuming them.

Next Richard had Naku demonstrate her ability to change the shape of her melon, which enables her to focus her echolocation in varying ways. She moved her melon from side to side and from front to back, and when I put my hand on it, it was like a plastic bag full of jelly that I could easily manipulate. For the rest of my visit with Naku, I couldn't help but repeatedly place my hand on her melon to feel it and watch it jiggle.

Naku then positioned herself so I could touch her back—her skin felt like a hard-boiled egg fresh from its shell—and look into her gray and pink blowhole. Then she rolled over with her belly pointed skyward and showed off what Richard called her rails, two large columns of blubber that run alongside her body from her naval to her anus. "Belugas are the only species of cetacean that distributes blubber in a cylindrical way," he told me later. "The heavier they get, the larger the rails get along their sides. Males tend to have more prominent rails than females, suggesting that it may be a secondary sex characteristic. However, the prominence of the rails varies between populations and seasons, making it an unreliable way to distinguish males from females."

I got to touch Naku's dorsal ridge, which is just an extra-thick layer of skin, and her tail flukes, giving me a close-up view of the leaf-veins that I saw from afar earlier. Richard said the veins are actually blood vessels near the surface of her tail that allow belugas to expel heat when they are overheated. Lastly, Naku demonstrated some vocalizations. Richard told me to use four fingers to tickle her under the chin, which prompted Naku to produce a startlingly loud and rapid clicking noise. A four-finger tickle to the roof of her mouth resulted in a soft kitten purr that could not have been more adorable from such a large animal. And drawing one finger across the side of her melon generated a loud fart-like noise, which was an entertaining way to wrap up my visit.

Richard used my time in the water as a training session for Naku, providing positive reinforcement—food—to reward her each time she exhibited the proper behavior. Yet even when he signaled to Naku that there was no more food left, the animal continued to respond to commands and allow Richard to repeatedly slap her tongue to make a loud popping noise. If I didn't know better, I would have guessed that Naku enjoyed the experience as much as I did.

TWELVE

GAINING GROUND

AFTER MORE THAN A WEEK OF WATCHING NARWHALS EASILY avoid our nets, the researchers were frustrated and the rest of us hardly bothered to go on alert when the whales approached. During some down time, I strolled the length of the beach and found an old, lichen-covered seal skull that looked like a unique art piece, which I kept for my skull collection (though I learned later that removing it may have been illegal). As remote as our location appeared, the skull was one sign among many that Inuit hunters have used the site regularly over the years. I also found plenty of bullet shells hidden among the cobble-stones, the skin and innards of a recently killed ringed seal that left a conspicuous oily spot on the ground, and a lone caribou antler that showed distinct evidence that it had been sawed from the animal's skull. As I glanced from the beach to the water, a male king eider surfaced, a handsome sea duck whose black-and-white body contrasted sharply with his distinctive clown-like mask in shades of orange, gray-blue and mint green. It's one of my favorite birds, and seldom have I had a better view.

The beach also had a variety of dazzling native plants. Dwarf fireweed, a stunning purple flower called *paunnat* by the Inuit, was abundant around camp. It grows best in disturbed soil and its wind-borne seeds are easily dispersed, so it is a common species along Nunavut roadsides, and I saw it everywhere in Pond Inlet and Qaa-naaq. Yellow-petaled Arctic poppies, one of the tallest plants in the region, and moss campion, which grows in a rounded cushion cov-ered in pink flowers, were also common. But my favorite was still the nodding bladder campion, an inconspicuous perennial herb with the Japanese lantern-like flower that I had first seen in my previous trip to the region.

The next time the narwhals marched by, they made all sorts of unexpected noises at the surface. Their breathing sounds were usually obvious, but their exhalations were hardly noticeable over the bellowing commotion that sounded like nose blowing or grunting. One of the Inuit members of our team, James Simonee, said that he can often tell from a distance that a big male narwhal is approaching by its distinctive grunt. But there were also an interesting variety of clicks, whistles, and other sounds that I thought the animals only made under water. For some reason, though, they were especially talkative that day.

A subset, we believed, of the larger herd, this group exhibited a wider variation in body color than I had noticed previously. There were an unusually large number of pinkish gray calves, noticeable both for their coloration and for their need to surface to breathe more often than the adults. One group of about twelve narwhals swimming together appeared to include five of the little sausages. But there were also a greater number of older pale gray animals mixed in with the usual charcoal gray adults.

The appearance of these animals prompted an interesting bit of discussion among the scientists, who had different opinions about which noises were vocalizations and which were sounds made from their blowholes. While all seemed to agree that the squeaks and clicks were vocalizations, there was no agreement about the bellows, which some thought was the breathing of a whale with the flu while others guessed it was another vocal variation. The researchers also raised unanswered questions about whether the narwhals were using their eyesight to see and avoid our net or whether they detected it with their echolocation. If they use their eyesight, then might they also see our encampment on the shores when they surface and should we, therefore, move the net further from camp so our colorful tents and constant activity don't frighten them away? No one had any clear answers. When I raised a question about the purpose of the tusking behavior we saw the day before, they just shrugged their shoulders as if to say, "That's a question we may never answer."

Everyone agreed, however, that the icebergs traveling down the fjord were headed straight for the net. But they were moving so slowly that it was hard to tell whether they were going to arrive within an hour or a day. I had the 3 a.m. watch duty the following day, so I tried to go to sleep early, but shortly after midnight I was awakened by a loud and alarming yell, "Grab the boats!" I glanced out my tent to see six of the team struggling to untie the Zodiacs. One of the icebergs—a long, low, flat one, perhaps forty feet wide and 100 feet long—was sliding by the near edge of the net and in direct line with the boats. I threw on my clothes and ran to help, just in time to pull one Zodiac from the water while the other was driven out of harm's way. We ran to untie the rope connecting the net to shore and handed it off to the boatmen so they could retrieve the net. An SUV-sized iceberg barreling down the waterway almost beat them to it, and the Zodiac collided briefly with the berg but managed to maneuver the net out of harm's way before spinning around and letting the ice continue on its way. Those in the boat detached the net from the anchor buoy and raced to shore, where-upon the rest of us hauled the now very heavy vessel out of the water just in time for a long, three-legged iceberg to cruise through where the net had just been. As it did so, it ran over the anchor buoy, which Orr said made it look like a giant squid consuming its prey. Moments later, the buoy reappeared from beneath the iceberg after having been dragged about fifty yards. It was an exciting and exhausting few min-utes, but a welcome respite from the boredom of the previous week. I only wished that it hadn't happened during the few hours I had hoped to sleep that night.

We didn't put the net back out until the next morning, when Orr decided to try a new strategy—using the Zodiacs to chase the approaching narwhals into the nets. When the whales returned to the area the next time and avoided the net once again, camp helpers Frank McCann and Joshua Idlout jumped into one of the boats and raced out to try to redirect the animals back toward the net. Revving the engine, they zoomed around in circles and tried everything they could to herd the animals in the right direction. It was obvious from the start that the narwhals were immediately confused, turning around and around, and uncertain of where to swim to get out of the boat's way, which was the objective—to get them off guard so they were no longer paying attention to the net. Rather than swimming along together, as they had been doing every day for weeks, they took off in different direc-tions, turning on a dime, reversing themselves and diving out of the way of the boat. I was watching the animals through my binoculars and reporting to Orr where the closest groupings were surfacing, and Orr told the boatmen via radio where to go next. In the confusion, several groups of narwhals came so close to the nets they barely avoided becoming entangled. For thirty minutes, the whales swam back and forth, with the Zodiac zipping around like an out-of-control race car. Finally, while the boat was far to the south, a group of narwhal stragglers, appearing oblivious to the noise and confusion, slowly glided toward the net, breathing comfortably at the surface as if they were taking a slow-motion stroll around the neighborhood. We looked on in amazement, tightened our gear one last time, and prepared to race into action. When the narwhals looked to be too close to the net to avoid it, they all casually dived and resurfaced just a few feet on the other side, not even jostling the buoys. I was amazed at their skill, yet dejected at our continuing failure, knowing that our best chance of catching a live narwhal may have just come and gone.

The Zodiac continued chasing the whales for a few more minutes, but by then the animals had regrouped, moved to the far shore, and headed back to where they came from. It was the most exciting thing to happen on my watch during the entire trip, and I finally felt like I had been somewhat useful to the researchers. But in the end, it was another fruitless effort.

By then it was time for me to leave. It had been both a fascinating experience and a frustrating one, since we had been unable to catch a whale while I was there. But while the rest of the team was staying for another week, my time in the Arctic was up. At midnight, with less than twenty-minutes notice to take down my tent and pack up my gear, the Inuit boatmen took me on a six-hour voyage on a small open vessel through frightening swells and driving rain back to Pond Inlet, where they planned to retrieve a different net that the researchers thought might be their last chance of catching a narwhal.

×   ×   ×

While the research team continued its efforts to reveal new insights into the life cycle of the narwhal, several young graduate students from Canada, Denmark and the United States were trying to answer other questions about the whale.

Whenever I asked about "the girls on the rocks," everyone knew who I meant. That's even how the two students referred to themselves: Marianne Marcoux, a doctoral student at McGill University, and

Marie Auger-Méthé, who was a master's degree candidate at Dalhousie University when I first met her but who was pursuing a PhD at the University of Alberta when I caught up with her later. They spent the month of August for three consecutive summers living on the rocky promontory known as Bruce Head, the steep point of land that formed the gateway to Koluktoo Bay, a three-hour boat ride from Pond Inlet. On my first visit to the region, we passed by them several times and they smiled and waved with enthusiasm, apparently happy for human contact. We even stopped there for a few hours and enjoyed some of our best narwhal watching of the trip, and I relished chatting with them at length about their research and their experiences living alone for weeks at a time in challenging conditions.

Over the next two years, I thought of them often. We emailed several times and I talked to each of them on the telephone, and soon I felt that I had come to know them, though what I knew hardly scratched the surface of their lives and their personalities. Marcoux initially appeared to be the more advanced student, but since French was her first language and she didn't seem entirely confident with her English, she often deferred to Auger-Méthé when the three of us were together. They were similar in height and build and hair color, and I'm sure I would have difficulty telling them apart if I saw them again. Until, that is, I heard them speak. Their accents were distinctive. I got the impression that they thought of their experiences on Bruce Head as an exciting adventure that also happened to be contributing to the scientific understanding of narwhals, sort of the way that I thought of my travels in search of narwhals. They pitched their tents high up on the rocks where they had spectacular views of the bay, the tundra, the icebergs, and the infrequently setting sun. They faced heavy wind and rain—their first year it rained for three weeks straight—and near-freezing temperatures, and they often depended on the kindness of the periodic passers-by for drinking water. They couldn't go more than fifty yards away from their campsite because they had no boat and there was no safe way of getting from their cliff to the rest of the mainland. Their only regular contact with people were the hunters who occasionally joined them on the rocks, knowing that Bruce Head was not only the best place to observe narwhals passing by close to land but also the best place to hunt them.

Every day for weeks at a time they stood watch around the clock for narwhals to glide by the peninsula. Sometimes they went for two days without a sighting. Other times there was constant activity for six hours at a time—often in the middle of the night—when they attempted to photograph the whales, note their behaviors, and record their vocalizations. Regardless of how many whales were about, one of the researchers was on the lookout twenty-four hours per day, watching for narwhals and watching out for polar bears.

"We were used to working long hours," said Auger-Méthé, "but this was particularly challenging because of the really long days. The really long sunlight meant that we were on standby all the time. Often narwhals would arrive at 2 a.m., and that made for a weird sleeping schedule. At the end of the season I was really wiped out."

Several years earlier, as Auger-Méthé was completing her bachelor's degree and Marcoux was finishing her master's, they often chatted about what fun research project they could collaborate on for their next academic degrees. Both had grown up in cities—Marcoux in Quebec and Auger-Méthé in Montreal—and neither had any particular interest in narwhals. But their advisor at Dalhousie University, Hal Whitehead, had theorized about a concept he called cultural hitchhiking, in which some species evolved based more on their culture than on their genetic makeup. While the question of whether animals even have a culture is somewhat controversial, Whitehead took that as a given and proceeded to attempt to identify which cetaceans were cultural hitchhikers. He wasn't sure about narwhals, because little is known about their social behavior. So the two friends imagined themselves finding the answer, never expecting that they would actually make their way to the Arctic and see narwhals.

In order to learn about narwhal culture—their social interactions, the behaviors that are shared among group members and those that are learned from other members of the group—it was important to be able to identify individual animals. The way that biologists identify individuals of most whale species is by distinct characteristics of their dorsal fins or their flukes. But narwhals don't have dorsal fins, and they seldom raise their flukes out of the water enough to see if they have distinct markings. So Auger-Méthé decided that her research would focus on establishing a method of identifying individual narwhals and creating a photo identification catalog of the narwhals that spend the summer in Koluktoo Bay.

"Every scientific study is improved if you can recognize individuals," she said. "It's useful in population estimates; you can look at residency time—how long they stay in the bay and if they come back every year; you can learn about social structure—who's hanging out with who, are they forming stable groups or mixing; behavioral studies, like who's caring for the young. You can only do that if you can recognize individuals. It's an important step to answering lots of different questions."

So Auger-Méthé, who became fascinated with the Arctic after her mother worked on documentary films about the region, stood for hour after hour and day after day on the rocks closest to the water taking photographs of every narwhal that passed by within a certain distance. Back at school, through months of analyzing thousands of photos, she tried to identify individual whales by the pattern of nicks and notches on their dorsal ridges, pigmentation patterns on their bodies, scars from bullets, and various scratches. She categorized all of the markings and looked at how prevalent they were in the population and whether they changed over time. She found that while pigmentation patterns change from year to year, they could still be useful in short-term studies. But the most useful markings for identifying individual narwhals were the nicks and notches on their dorsal ridge. Although no one is certain how they occur, Auger-Méthé says they seem to accumulate more and more of them over time. She likens them to wrinkles on human faces, which occur as we age due to wear and tear and never disappear.

She created a computer program to automate the process of analyzing photos and matching them to known individuals, hoping that other scientists will use it in future narwhal studies. She envisions that members of the Inuit communities may use her method to monitor their local narwhal populations. While Auger-Méthé has moved on to studying polar bears for her doctoral dissertation, she still has a soft spot in her heart for narwhals.

"They're just so weird," she said with a sweet laugh. "They swim upside down; that's so ridiculous! They have a tooth growing out of their face and no other teeth in their mouth; that's ridiculous to me. The sound they make when they vibrate their blowhole and they take a breath sounds like farts. They're a huge herd of farting animals," she continued, laughing louder now. "I had this romantic idea of narwhals slowly swimming wonderfully by, and when we saw our first narwhal and noticed their farting noise, I realized that they aren't nearly as romantic as I thought they'd be. They crack me up! They are so not boring. They are the reverse of everything."

When she reflects on her time at Bruce Head, Auger-Méthé speaks warmly of an Inuit guide named Luke who stood guard during the researchers' third season in the Arctic. She said that she and Marcoux hired a local guide each season to help protect them from polar bears— they only saw two bears during their three months of field work—and to contribute to the local economy.

"Luke was just a teenager, so it could have been a horrible experience, but the time I spent with him I really cherished. He'd sit with me on watch, we'd play cards, and he taught me some Inuktituit (the local language). I had a lot of fun with him."

Perhaps the greatest difficulty Auger-Méthé faced was the emotional strain of watching the hunters kill her study subjects right in front of her.

"It was emotionally challenging to have to deal with the hunting," she said. "Seeing the animals that you are studying being killed is really hard, especially when I was trying to take a photo of a narwhal that was then killed. My goal was to re-identify narwhals, but many of the photos I took were of narwhals that were killed right after I took the photo, so I knew I wasn't going to see that animal again.

"Coming from the city, we don't kill our own meat, so we're not used to seeing animals dying," she continued. "Maybe it was good for me to learn, because it's not something that my generation is used to. But seeing them die was pretty hard, and it doesn't happen very quickly. Seeing one sink was also hard to watch, because you know that it got killed and no one was going to use it. It got killed for nothing."

I felt for her, knowing that I would be equally disturbed by the experience, even after having spent time in a narwhal hunting camp and seeing a narwhal harpooned and hauled to shore. Watching as her research subjects were killed before her eyes, with little opportunity to even state her objection for fear of alienating the Inuit community on which she was dependant, must have been devastating.

× × ×

For Marianne Marcoux, the greatest challenge to studying narwhals is their location in the Arctic. There are only two months of the year when they are close enough to shore to study, and even then, the environmental conditions make it difficult to collect the necessary data. Coupled with the fact that so little is known about the whales, she knew very early on that she might have bitten off more than she could chew.

Her aim was to understand narwhal society, with the ultimate goal of solving the question of whether they are cultural hitchhikers. It would have been useful if Auger-Méthé had completed her photo identification project before Marcoux started her research, because being able to identify individual narwhals is crucial to the study of social behaviors and group dynamics, of mother-calf interactions, and male dominance relationships. It would have also been useful to have access to a boat to follow particular groups of animals, but that wasn't in the cards either. "If we could follow them, we would get more data about different behaviors," she said. "All we really got from Bruce Head was a snapshot."

But even that snapshot provided insights never before gleaned by other narwhal researchers. Through 12,650 observations of narwhals in 2007 and 2008, she found that most traveled in clusters of three or four animals, though some clusters numbered as many as twenty-five individuals. Clusters were almost always sexually segregated, with males traveling exclusively with other males and females grouping with other females and their calves. In many cases, the clusters were segregated by age as well, with older whales interacting most frequently with those of similar ages, much like human societies. The clusters were part of mixed herds that consisted of up to 642 clusters, though most herds contained about twenty clusters.

"Our most significant conclusion," Marcoux said, "was how much they were segregated by sex and age. Maybe they form groups when they're young and they stay together and grow together."

It was difficult for the girls on the rocks to draw conclusions about other behaviors. Almost all of the narwhals they observed were simply traveling into or out of Koluktoo Bay, so there weren't many social behaviors being exhibited. Nonetheless, Marcoux thinks fondly of her time at Bruce Head.

"The first herd of the year is always good, though the whole thing is pretty exciting. What I really liked was on the calm days and you hear the narwhals coming, first a few and then hundreds. It was overwhelming. On those calm days you can hear them very, very well," she said. "The most magical moments were when you could see the males with their tusks, when they were socializing and raising their tusks together."

Marcoux laughed as she told me about a dog that one of the Inuit guides had with him during their stay on Bruce Head.

"They would bring a dog along to alert us to polar bears, but the entire first year we never heard it bark," she said. "Sometimes we wondered if a polar bear had arrived and eaten the dog first."

While Marcoux and Auger-Méthé were camped at the entrance to Koluktoo Bay, they regularly tried to record underwater vocalizations of narwhals to study their acoustics. On posts jammed between rocks, they strung a long cable from their tents at the top of the overlook to the water's edge. They were hoping to learn if they could identify the sounds narwhals made in different behavioral contexts. And they believe they did.

"They used different whistles for different behaviors," Marcoux said. "Whether the narwhals were traveling at good speed or resting or socializing, we could measure the frequency and duration of the whistles, and those characteristics were different for each behavior."

She noted that, in addition to the whistles, narwhals also exhibit what she described as a pulse call—a series of clicks—and while the pulses did not differ from behavior to behavior, each herd seemed to use a different pulse. Marcoux said it could be that each herd has its own dialect.

× × ×

Communication among whales is a subject of great interest to many scientists. The long, haunting songs of humpback whales, for instance, have been studied for decades, and recordings of these calls have been incorporated into popular music, film scores, and soothing relaxation compilations. But little is known about the vocalizations by narwhals. When I listened to the sounds they made on my first visit to the High Arctic, they sounded scattered and unorganized, like conversations in a noisy auditorium. But a few researchers have taken a closer look at this cacophony and uncovered some interesting clues.

Like most toothed whales, narwhals have developed a sophisticated, sound-based means of finding food and exploring their environment. By controlling the passage of air between chambers near their blowholes, they can produce a variety of whistles, clicks, and knocking sounds that some scientists believe are then reflected off the front of their skulls or redirected by their melon. A fast series of clicks—sometimes called "click trains"—is used for detecting obstacles at short ranges or for the echolocation of prey. Produced at frequencies up to 160 kilohertz and more than 200 decibels, some of these click trains may be strong enough to disable or disorient prey. Whistles, trumpeting, and squeaky door sounds lasting up to several seconds range from 48 to 500 kilohertz and are probably used for social communication, with males having a larger vocal repertoire than females.

Science radio reporter Ari Daniel Shapiro sent me several audio files he recorded of individual narwhals vocalizing, and they could only be described as bizarre. Had I not known they came from a whale, I would have guessed it was a practice session for the sound effects guy from my favorite National Public Radio program, *A Prairie Home Companion*. One narwhal alternated between what sounded to me like a creaky rocking chair and a whistled squeak, while the other made noises that clearly corresponded to those made by the first narwhal but were definitely different. The second animal's vocalizations were more high-pitched and the whistled squeak sounded like air being released from a balloon or a teenager making bathroom noises by blowing air through his clenched fist. The apparent distinctions between similar vocalizations made by individual narwhals are what caught Shapiro's attention.

While studying for his doctorate at the Woods Hole Oceanographic Institution, he made the recordings during a summer 2004 visit to Admiralty Inlet on the north end of Baffin Island with a team of narwhal researchers led by Canada's Jack Orr and Denmark's Rune Dietz. The scientists attached two types of tags to several narwhals—a surgically implanted satellite device for long-term tracking of the animals, and a digital archive tag called a D-tag that attaches to the whale's back with a suction cup and records the whale's movements and any sounds it makes or hears in the water.

About the size of a cell phone, the D-tag had been invented just a few years earlier by Woods Hole engineers to study endangered North

Atlantic right whales to learn why they are hit by ships so often. The tag is essentially a miniaturized computer that records high-quality sound using a built-in hydrophone. It also contains a digital compass, a pressure sensor, and an orientation sensor to reliably measure the whale's pitch and roll. In addition to the right whales, it has been deployed on manatees in Belize, pilot whales off the Canary Islands, sperm whales in the Gulf of Mexico, and humpbacks off Australia. But until Shapiro's narwhal project, it had never been deployed in cold water. As a result, he had been concerned about the release mechanism, the part that enabled the tag to detach from the whale and drift to the surface, which had been engineered for use in warm water and never tested for Arctic conditions.

"Usually when using these tags you follow the animal in a boat, and when the tag falls off it floats to the surface and you just grab it," explained Shapiro. "But with the narwhals, we were camping on the shore, and the weather and wind made it too dangerous to follow the animal. So we just waited for it to come off. We weren't sure when it was going to happen and if we would be able to retrieve it, so there was a lot of nail biting between when we put the tags out and when we got them back."

Shapiro deployed D-tags on three narwhals. The first remained attached for about two-and-a-half hours and was quickly retrieved, while the second collected twelve hours of data but took several days to recover. The third was never found. After analyzing the recordings, it was clear to him, just as it was to me, that each of the animals produced two different kinds of sounds but they sounded a little different from each other. He suggested that this may be preliminary evidence that narwhals produce what he called "signature vocalizations."

"It means there is content—something related to pitch or timing or how long it is—that is consistently variable between individuals," he explained. "There is something unique at the individual level. The frequency was the same, the duration was the same, the structure of the calls was the same, but it differs between individuals. And if I can tell them apart, then the narwhals can tell them apart." He said that the next step was to determine the vocalization repertoire of as many narwhals as possible to see if it sheds light on whether the whales are using the sounds to tell one another apart and in what context they use them.

I asked Shapiro if this discovery was akin to narwhals having individual voices, like when humans can recognize individuals on the phone after hearing them speak just a couple of words.

"It's not the same thing as a voice," he said. "If we each say the word narwhal, someone could tell the difference between us because of the differences in our vocal tracts. In marine mammals it's different, more like a completely different vocalization. They each produce a whistle, but it's a different kind of whistle. It's the same type of vocalization, but it has a different formulation or structure. It's hard to say what they're using them for."

What he can say for certain, however, is that his acoustic recordings of narwhal vocalizations provide a rare glimpse into their communication patterns.

"When other vocalization systems have been studied, it has opened up a deeper view of the social structure of the animals, how they relate to each other, how they communicate about relationships, feeding, and other issues. It's a portal into figuring out their social dynamics and their communications dynamics. And it helps us understand their behavioral ecology," Shapiro said. "If we get a sense of why marine mammals use sound and how they use sound, then the question of noise pollution naturally comes up. If they rely on sound, then ship noises and other man-made noises probably have an effect on them. How much noise pollution should we allow in their environment?"

During the same research expedition, the scientists deployed a Crittercam on a narwhal to gain insight into the underwater behavior of the animals. Developed by the National Geographic Society, the special video camera was strapped to the back of a narwhal, near its head, so the camera could see what the whale was seeing. Much to the scientists' surprise, the results showed that narwhals spend a considerable amount of time swimming upside down when they are near the seafloor, the first time this behavior had been documented.

"As it went down, it was rolling and corkscrewing, and when it got to the bottom, it spent a lot of time upside down," Shapiro said. "There's lots of speculation as to why that might be. Rolling behavior has been documented for sperm whales, which is believed to be for prey detection and capture."

That's what Rune Dietz thinks narwhals use their upside down swimming for as well, though there are plenty of other theories being bantered about by scientists. Dietz said that narwhals direct their echolocation sonar through the melon on their forehead to detect prey, and the direction of the sonar must be oriented toward the bottom where their prey is found. "It's like the directional sonar in a boat that is placed in the stern pointing towards the bottom at an angle," he said. "Maybe it's not ideal when you want to grab prey, but I suspect they can rapidly turn around." If a narwhal wanted to point its melon at the bottom while swimming normally, and if it was a male with a tusk, the tusk would likely get in the way or strike the bottom, making it more difficult to catch its prey, he added.

The D-tag the scientists attached to the narwhal as it carried the Crittercam through the water also provided data on the whale's "roll dynamics," which indicated that it rotated both clockwise and counterclockwise, suggesting that the direction of the tusk's spiral plays no role in the direction it rotates under water. Why the tusk spirals at all still remains a mystery.

× × ×

After dropping me off in Pond Inlet, the Inuit boatmen returned to the Tremblay Sound research camp with a new net, and just thirty minutes after it was deployed, it caught a narwhal. It was a female with a calf nearby, and she was just barely entangled in the segment of netting farthest from shore. In the researchers' cautious efforts to disentangle her, she escaped. But it was a clear sign that their chances of catching narwhals during the final few days in camp had increased exponentially with the arrival of the new net (and, I hope only coincidentally, by my departure).

Before I left, Orr told me that he was certain that there was something about the structure or design of the first net that made it visible to the narwhals. "That net has caught whales before, but every year is different," he said. "These whales have memories and they do see the world around them, so if we're seeing the same whales coming in and out of here, they know there's a potential danger here, so they're avoiding it. . . . [Tagging whales] isn't an exact science. It is something that changes with every location, and even different stocks of whales may be more [aware] at certain times of year."

When I talked to him a few weeks after he returned to his Winnipeg office, he said simply, "That's how it goes sometimes. One little thing can change the whole success of the trip. Whether it was the color or the mesh size or that it hung a little deeper in the water, I don't know, but changing nets worked."

It sure did. At 3:15 a.m., just a few hours after the first narwhal escaped from the net, two males with tusks hit the net at the same time. But it was an inauspicious start for the scientists. Most of the researchers were asleep, and with rough seas and dark, overcast skies, even those on watch hadn't noticed any whales in the vicinity. When the team was roused and sprang into action, they had difficulty starting the Zodiac motors, and one animal's tail became extremely entangled, so pulling the whales to shore took longer than expected. It turned out that both narwhals were old males with broken tusks—one tusk was severed in half and extended just three feet in length, while the other was a bit over five feet long and missing just its tip. Both animals were fifteen feet long with flukes nearly three-and-a-half-feet wide, and one had a six-inch scar from an old bullet wound that was colonized by a mass of sea lice, which feed on exposed flesh. In photos of the lice the researchers shared with me later, it appeared as if dozens of caramel colored beetles or engorged ticks were surrounded by dozens of juveniles and hundreds of tiny larval-stage lice that could easily have been mistaken for golden caviar. Orr said that sea lice are common on the wounds of most marine mammals, and they are almost always found on male narwhals at the base of their tusks. Their hooked legs sometimes even leave tracks and trails across the whales' skin.

Meanwhile, Martin Nweeia's introduction of highly saline water into the tusks of the captured whales resulted in contrasting results. He said that the whale with the nearly-complete tusk responded like male narwhals did in previous studies, by raising its heart rate. The animal with the severed tusk did not respond in kind, which is what Nweeia said he anticipated, since the nerve mechanism in the pulp of the tusk had been destroyed when the tusk was broken. According to Nweeia, these results back up his belief that the tusk is a sensory organ, though none of the biologists I spoke with were convinced by this new information.

While Nweeia conducted his tests, veterinarian Sandie Black began her examination, taking a blood sample from a vein on the underside of one whale's tail. Then she turned around and collected a culture from its blowhole so scientists can learn about the bacteria in its respiratory system. She also assessed the animal's respiratory rate by watching and listening to its blowhole, and conducted a physical exam of the animal, looking for any superficial wounds that might be associated with the capture, noting any healed or healing wounds on the body, and checked its eyes. "It's not a particularly thorough physical examination in the way you look at a terrestrial animal," she told me. "[I can only see] maybe a quarter of the animal and any other bits and pieces that are visible as we're working. . . . I'll get the heart rate and respiratory rate three or four times while we're handling the whale. As soon as they're getting close to the end of the tagging, I'm back to the tail for another blood sample so I've got the comparison of how the animal is doing from the beginning of the handling to the end."

Exactly one day after the two male narwhals were captured in the nets and processed by the researchers, a thirteen-foot-long female became entangled. She was part of a group of narwhal cows and calves that waited nearby during the thirty-five minutes it took to process the captured whale. Five hours later, a lone female swam into the net, and thirty-six hours after that, another lone female followed suit. Both of these animals were just a few inches shy of thirteen feet in length, and one apparently had a near-deadly encounter with a polar bear sometime in her past. She had long parallel white scars extending three feet along her sides like she had been scratched by a bear's claws, and she had what looked to be bite marks around her blowhole. Jack Orr suspected that the whale had been in an ice entrapment, which provided the bear with easy access to her. "He would have launched himself onto the whale's back and grabbed on with his teeth around the blowhole," he surmised. "As she went down, his claws raked her behind the pectoral flippers. The bear had all four feet and his face into that whale, but I guess she had enough mobility to dive and he let go." The scars were well healed, he added, but Orr guessed that the attack may have occurred as recently as the previous winter.

Orr was pleased that the researchers were able to capture and process five narwhals during the three-week project, especially considering how they had been skunked during the first two weeks. And he was looking forward to monitoring their movements and behaviors to see what he could learn from them. He was intrigued that, according to the satellite tag data, two months after their capture the two males were still traveling side-by-side despite being several hundred miles south along the east coast of Baffin Island. The three females were in various locations in Navy Board Inlet and Eclipse Sound, still within fifty miles of where the researchers first saw them in Tremblay Sound. By late November they were all in the middle of Baffin Bay at the edge of the expanding pack ice, and one eventually made it all the way to Disko Bay on the central coast of West Greenland.

While it's easy to understand how tagging data can provide insights into the migratory patterns of whales, one of the more interesting results from narwhals Orr tagged in previous years offered greater understanding of the animals' diving behavior. Coupling minute-to-minute location data with data about how deep individual whales dive—both gathered from the satellite tags—and merging that with geographic information about water depth and ice cover, he discovered that narwhals appear to exhibit two different diving behaviors.

As Orr described it to me, during the winter months when narwhals are living in dense ice, they typically use a V dive for feeding and a U dive for finding their next breathing hole. During a V dive, the animals dive straight down to feed at a certain location where they know they can find food—perhaps it's on the bottom where halibut are found, or it may be higher in the water column if they know squid are in the area—and then they swim straight back up to the surface to a different known breathing hole.

"Whereas with a U dive," Orr said, "they're going down and then going horizontal for a while and then they'll start their dive up toward the surface again. What we presume is that they're getting to a certain depth and then using their echolocation to find their next breathing hole. . . . These animals are living in such dense ice over the winter months that they're coming up in breathing holes that aren't really even a hole that you would see at the surface. That's a behavior that they've adapted to navigate in dense pack ice."

THIRTEEN

LOOKING AHEAD

ON MY VISIT TO THE CANADIAN ARCTIC TO JOIN JACK ORR'S research team, I arrived in Pond Inlet just as a parade of twenty-four boats was returning to the tiny harbor filled with joyous residents waving flags, shooting guns into the air, and laughing and hollering in celebration of their first successful bowhead whale hunt in several years. While narwhals are a more available and dependable and accessible food source for Inuit families, the killing of just one adult bowhead would provide enough muktuk for the entire community for many months and keep them warm and happy at least until the end of the year. When they are allowed to be harvested, bowheads are just as important to the local culture as are narwhals. Canadian Inuit communities must apply for a permit to hunt bowheads, and they are only granted permission every few years on a rotating basis, so the 2010 hunt was a day the residents would long remember. As I learned in speaking with the Inuit boatmen who took me from Tremblay Sound to Pond Inlet, it's a major undertaking to capture a bowhead, involving nets, harpoons, guns, and several boats to haul the animal to shore for flensing. That's why so many members of the community traveled to the distant fjords west of town to participate.

Bowheads are massive animals—as long as sixty-five feet and weighing more than fifty tons, with a blubber layer eighteen inches thick that makes it well adapted for cold temperatures. Their brown eyes, according to Barry Lopez, are "the size of an ox's" but are nearly lost in their huge head, which they use to break through six-foot thick ice to reach the surface to breathe. Their slow and predictable movements and long periods spent at the surface, however, made them an easy target for commercial whale hunters. About 30,000 bowheads were killed in Baffin Bay over a period of about 250 years at the peak of the whaling era. By the late nineteenth century, the population had been decimated and the species was nearing extinction.

With a more rotund shape than other large whales and a black-and-white pattern on their head and tail, bowhead whales are the only baleen whale to spend their entire lives in the Arctic. They feed on several species of copepods—microscopic zooplankton found in abundance in the region—by taking in a mouthful of sea water and using their tongue to push that water out through the 300 baleen plates that hang six to ten feet from their upper jaw to filter out the tiny organisms. They have a circumpolar distribution, but the animals found in West Greenland and eastern Canada have not been known to leave this area or mingle with bowheads in the waters of Alaska, Russia, or Spitsbergen (the largest island in Norway's Svalbard archipelago). Or at least that was the case until 2011, when Mads Peter Heide-Jørgensen tracked bowheads from the Alaska and Greenland populations crossing through the ice-free Northwest Passage and lingering together in the same area. Their more typical migration, however, is rather unusual in itself.

According to Kristin Laidre, who has studied bowheads with Heide-Jørgensen in Baffin Bay for nearly a decade and who has used satellite tags to track their movements, the largest congregation of bowhead whales in Greenland appears in spring in Disko Bay on the central west coast. Perhaps as many as 1,200 bowheads can be found there, mostly adult females, and they undertake an annual migration that takes most of them all the way around Baffin Island. In the summer, they depart from Disko Bay and work their way north and east toward the east coast of Baffin Island and up through Lancaster Sound. They then move south into Prince Regent Inlet, Repulse Bay, and eventually down into the Hudson Straits, where they spend the winter before following the retreating ice to Disko Bay again in March and April. Most of the mothers and calves are believed to spend the entire year in Canadian waters, especially in and around the shallow waters of Foxe Basin on the west side of Baffin Island, perhaps to protect the young whales from the aggressive sexual activity of the breeding adults.

Laidre said that there are probably about 15,000 bowheads in the waters of West Greenland and the eastern Canadian Arctic, and the numbers have grown considerably in recent years. "In West Greenland, they surveyed hundreds of thousands of kilometers in the 1980s and saw maybe one bowhead. It's only been in the past ten or fifteen years that there has been an explosion of bowheads up there. And it's not really well understood," she said. The whales are no longer commercially hunted, so their numbers are increasing, but they reproduce very slowly. It is believed that their recent abundance around Disko Bay is a combination of a recovery in the population and animals that are returning to historical areas, but there is likely an influx of bowheads from other areas as well. "There are lots of different whale species that come there in summer to feed," Laidre said. "It may be because the ice is breaking up earlier, the bowhead whales can move in to feed earlier and have longer feeding opportunities."

While the loss of sea ice and potential changes to the marine ecosystem as a result of global warming will likely have similar negative effects on bowheads as on narwhals and other Arctic marine life, so far it may actually be helping them. Because the spring algal blooms are occurring early, climate change is apparently extending the whale's feeding periods. Bowheads in the Bering Sea and Alaska are experiencing the same positive effect. "Down the road as things really change, it could be a problem, but for right now it's not such a bad thing for them," noted Laidre. "Their association with ice is much different than with narwhals. They're not relying on the leads like narwhals are, and bowheads can break a lot more ice."

On the other hand, bowheads may face a greater threat than narwhals from oil exploration and other industrial activity. Bowhead whales do not use echolocation like narwhals do to find food, communicate, and sense their surroundings. Like the rest of the world's baleen whales, they emit low frequency sounds to communicate instead, and the noise from offshore oil exploration and production, mining and increased shipping activities could disturb their ability to communicate.

Bowheads aren't immune to concerns about hunting, either. While commercial hunting ended decades ago, a limited subsistence hunt continues in Alaska and Canada, and Greenlandic hunters in the communities around Disko Bay were authorized by their government in 2010 to kill two bowheads per year. But despite the historic and present threat from hunting, bowheads may be the longest lived creatures on Earth, with some having been determined to be more than 200 years old. One indication of their age was revealed in Alaska in the 1990s when several bowheads were harvested and found to have stone and walrus tusk harpoon heads embedded in their bodies, both of which are believed to have been used prior to the 1880s. So if those bowheads were killed in the 1990s and had previously been hunted with hand-made harpoons in the mid- to late-1800s when they were most likely already adults, then they were probably more than 130 years old when they were killed and might well have lived for many more years.

× × ×

Determining the age of a whale and how long they may live is an especially challenging feat, yet this information is crucial for understanding issues related to their life cycle, ecology, population dynamics, and other questions. Scientists are only now beginning to learn how to age narwhals, thanks to Danish graduate student Eva Garde. The standard age-determination method for fish is to count the pairs of bands in their vertebrae (each pair typically represents one year of growth) or to weigh and measure their otolith, a dense bone in their inner ear that grows at a steady rate every year. Determining the age of toothed whales has been performed reliably since the 1980s by counting layers in their teeth. The method has been validated on many species by comparing the results to the teeth of captive animals of known age. But this method has not worked well on narwhals. Since narwhals have never been successfully kept in captivity and no one knows for sure the age of any individual narwhal, there is a great deal of uncertainty about the age estimates of narwhals. As a result, biologists cannot confidently answer many of the questions they have about various aspects of their life history, like their age at sexual maturity.

That's where Eva Garde comes in. She is a doctoral student at the Biological Institute of the University of Copenhagen and affiliated with the Greenland Institute for Natural Resources. In 2004 she was the first to use a new aging method on narwhals that analyzes, of all things, their eyeballs. Aspartic acid racemization is a technique that was developed in the 1970s to assess the age of fossils and marine sediments. A decade later, scientists experimented with it for age estimation of mammals, including whales and humans. It was by this method that scientists discovered in 1999 that bowhead whales probably live longer than any other mammal on Earth.

The new technique determines age by studying the changes in levels of aspartic acid, an amino acid found in the teeth and in the eye lenses. According to Garde, the amino acids normally incorporated into proteins in living organisms are of a form called L-enantiomers, which slowly—and at a constant rate—are converted to the D-enantiomer form. The ratio of the D form to the L form provides a marker for age. Garde collected eyeballs from seventy-five narwhals legally killed by Inuit hunters in West Greenland, stored them immediately at −20°C, and later removed the lenses for analysis.

What she found was a surprise. Twenty percent of the whales sampled were estimated to be more than 50 years old, and the oldest narwhal of the bunch was a female that lived for about 115 years. That's more than twice as old as previous estimates of the species' maximum longevity. Garde suggests that, since the narwhals she tested came from a heavily hunted population, it is likely that narwhals may well live even longer in populations that are less disturbed by hunting. She also thinks that the maximum age may increase as more specimens are examined.

As is typical of most scientific analyses, one outcome of Garde's research is a new question: Why are two of the longest living mammals on the planet—narwhals and bowheads—both whales that live in the Arctic? Early speculation centers on their close association with the sea ice and the importance of finding enough leads and cracks in the ice in winter in which to surface and breathe. Changes in sea ice cover caused by climate changes in the Arctic sometimes happen over just a few years time, making the whales highly vulnerable. That's one reason why the present climate changes are raising such concern for them. It may also be why the genetic diversity among narwhals throughout their range is as low as is usually seen in severely depleted whale stocks.

"One evolutionary strategy to overcome such sudden changes in climate could be high longevity, thus aging could evolve as part of an optimal life history," wrote Garde in a research paper. "The longevity of narwhals could therefore be seen as an adaptation to mitigate the population effects of drastic changes in climate."

Another outcome of Garde's analysis was her conclusion, based on her age data and the measurements of narwhal sexual organs, that female narwhals achieve sexual maturity at about six or seven years of age and physical maturity at nearly four meters in length on average, while males mature at age nine and four-and-one-half meters in length. These figures are comparable to those estimated by other scientists using different methods, which adds a level of confidence to Garde's research.

While it makes me recoil just thinking about the process Garde undertook to conduct her research—collecting eyeballs from freshly killed narwhals, removing their lenses in a laboratory, subjecting them to a chemical hydrolysis, and analyzing them using some high-tech equipment—it's also amazing what can be learned using these methods. I guess that's why she's the scientist and I'm just the one looking over her shoulder.

× × ×

The one thing I missed during my narwhal expeditions to the Arctic was a chance to take in what has been described by many as one of the most spectacular natural phenomena on Earth, the aurora borealis or the northern lights. I had seen pictures and videos of the spectacular light show it can produce, and I've spoken to several people who have experienced it, but none could adequately describe the mystical glow. It may be impossible to do so. A display that amazing just can't be put into words, they all said. So I was determined to see it for myself, to help me imagine what narwhals may see as they come to the surface to breathe during the midwinter darkness. Since all of my Arctic expeditions to see narwhals and meet with researchers and hunters took place during the summer months, when daylight lasted twenty-four hours, there was no possible chance that I could see the aurora during those trips. That meant I had to make a special winter trip, which I did just before completing this book.

The aurora borealis is interpreted in numerous ways by various cultures. According to Inuit legend, the aurora is the highest level of heaven to which one could hope to ascend, where there is no snow and where animals are easily caught. To get there, one must die in a hunt, be murdered, or die in childbirth. Some say it is caused by the spirits playing ball with a walrus head. In Greenland, the aurora is believed to come from the spirit of stillborn or murdered children playing ball with their afterbirth. In Alaska, they say that the light will come closer if you whistle at it, though others claim it will cut your head off if you whistle.

Regardless of the many beliefs associated with the aurora—including the common explanation that it is sunlight reflecting off of polar ice—science tells us that the phenomenon originates from solar flares erupting from the sun and sending energetic particles toward Earth. When the particles are deflected by the Earth's magnetic field, they travel around the planet until they enter the atmosphere over the poles and electrically stimulate gases, which glow from the attack. Oxygen produces a green glow, the most common color reported in the aurora, and sometimes red, while nitrogen generates shades of blue.

The scientific explanation of this stunning natural event pales in comparison to a first person observation of the phenomenon, and I had high hopes when we arrived in Iceland because a large solar flare had been reported just a few days before. But we were also in a period when the moon was full, which brightened the sky and was likely to dim any potential aurora display. Parked by the side of the snow-covered road leading into a national park less than an hour east of the capital city of Reykjavik, we watched and waited expectantly without a hint of the lights on our first night. Before we even reached our parking spot on our second attempt, however, we noticed an unusual band of cloud-like gases in an arc across the sky, like a pale belt across a bulging belly. It was an unnatural shape for a cloud and reminiscent of a colorless rainbow, so we pulled over to take a better look.

Almost immediately, colors began to emerge and compete with the light of the moon, though they were initially almost too faint to determine whether they were real or just our eyes playing tricks on us. The arc soon became filled with a pale greenish glow, like a translucent key lime pie. A second arc of almost imperceptible pale raspberry emerged for a moment above the first, suggesting a double rainbow, but it quickly disappeared. Soon the greenish glow began to move—not in a shimmering curtain like the way that many people describe, but a slow shift in depth from one end of the arc to the other. In a wave, the right edge intensified while the left became dull and the middle disappeared entirely. Then the middle reappeared while the left glowed brighter. It didn't happen quickly, but if I turned away for a moment, the brightness of the colors changed places when I looked back again. As the flames raged in different places from minute to minute, it became obvious why Barry Lopez called the aurora a ghost fire and why poet Robert Service wrote that the lights "writhed like a brood of angry snakes, hissing and sulphur pale." For a few moments, a vertical streak of pale blue-green shot upward from the horizon to the right of the arc and faded into Orion. A third, almost colorless, band formed briefly below the right side of the first, then they all merged into one in what appeared to be the beginning of the end.

After about twenty minutes, the show was over. The bands of light faded until just a fuzzy glowing ball remained, and then it, too, was extinguished. While the aurora made a shy showing and it didn't quite match the photos I've seen or the stories I've heard, I'm sure that if we had stayed longer or tried another day or if the moon hadn't been full, the display might have been brighter or more colorful or more like the proverbial curtain of dancing light. But I still got my money's worth. Nature's light show certainly lived up to its billing.

×   ×   ×

My long flight home from Iceland and the even longer trip from Greenland gave me plenty of time to process my time in the Arctic, especially the conflicting feelings I always seem to have about hunting. After I was no longer caught up in the thrill of the chase and feeling my competitive juices flowing during the narwhal hunt with Mads Ole Kristiansen near Qaanaaq, I was left with two indelible images from that day: the splash made by the whale when it was first struck by the harpoon, when its tail slapped the water out of pain or anger or fear or simply in a desperate effort to escape, and the arched back of the whale—discolored from blood flowing along it—as it surfaced repeatedly for air near the end of its life, when it needed more oxygen than usual to power its muscles away from harm's way. Those images, I'm sure, will never leave my brain. And I hope they don't, not because I hope those events never happen again (though I'll be happy never to watch it happen again), but because it reminds me of my own humanity and perspective as a wildlife conservationist. Yet I have concluded, for the moment, that narwhal hunting has its place in some Arctic communities, where employment options are extremely limited, whale meat is a staple food, and the practice of its hunting is carried out in a respectful, fair, and circumscribed manner as part of an ancient culture. I can't say that I will feel the same about the practice at another time or place, but for now, in Qaanaaq and in other

Arctic communities whose methods and thinking are similar, I don't see how I can object.

The images of the harpooned whale aren't the only narwhal experiences that stick with me. The midnight jousting session in Koluktoo Bay, the first time I had a clear look at a narwhal tusk, may have been the most mesmerizing moment of my life, and thinking back on it now more than three years later still gives me a jolt of excitement that I pray never diminishes. I'm still not sure whether it was the quiet, slow-motion activities of the whales, the spectacular rocky hillsides and iceberg-filled waters, or the unexpectedness of the experience after having already gone to bed—or, for that matter, the silliness of watching the narwhals joust while we stood around in our underwear in freezing temperatures—that makes that observation rise to the top in my mind.

There were plenty of other moments that I won't soon forget. The parade of narwhals traveling back and forth day after day in Tremblay Sound, skillfully avoiding the researchers' nets and providing daily doses of frustration while impressing us with their remarkable navigational abilities. The more determined parade of narwhals streaking close by our boat as we entered Eclipse Sound, caring not a whit for the threat we posed, for they were far more concerned about escaping from the pod of hungry killer whales chasing them. The honks and moos and twitters and whistles and creaky doors and other bizarre sounds the whales made as I listened in on their conversations with a hydrophone in the middle of Koluktoo Bay. The first time I touched a narwhal—still warm despite having been killed minutes earlier—its skin soft and leathery and firm all at once, and my first taste of its blubber and its meat soon afterwards, something I'll never say was tasty but which I'm pleased to have shared with the proud hunters.

I had weeks of adventures that make me more enamored of the narwhal than when I started this project. How it can thrive in its icy world and find food in the dark depths despite the tremendous pressures will always amaze me. Yet despite their great skill and flexibility and physiological adaptations that enable them to undertake their entire life cycle in conditions that few creatures can withstand, and despite what I've learned about their somewhat stable populations from the world's experts, I still worry about them. About changing weather patterns that may trick them into delaying their migrations just long enough to trap them in the unforgiving ice. About retreating sea ice that will almost certainly change their marine environment and shift where their prey resources are found. About increased competition from commercial fishing interests for food and increased competition for habitat from oil exploration and shipping companies. And, yes, about hunters seeking to fill the stomachs of their families and earn spending money at the expense of this legendary animal.

x   x   x

I'm not the only one who is worried about them. All of the biologists I spoke to expressed somewhat similar concerns, though none was concerned that they were heading toward extinction. I trust that these experts will know when it's time to be more than simply worried, and that they and their followers will sound the alarm if and when that time comes. And then it will be up to the rest of us to ensure that the proper steps are taken to protect this remarkable animal. In the meantime, I'll let the scientists have the last word in response to my final question to each of them: What does the future hold for narwhal populations?

"I'm not flagging the narwhal as a species that's going down the tubes," said Kristin Laidre. "I think that's unwise and incorrect. But that doesn't mean they're not a species that might be more sensitive or might be more influenced by some of these things more than others, and that's really a function of flexibility and adaptability. . . . Narwhals are kind of picky. They like to do things the way they do them. They're very predictable in their migrations, in terms of what they eat and when they eat. So if you think about vulnerability, a species that's a specialist, that has a niche carved out, is more vulnerable than something that's flexible and adaptable. . . . For that reason, we need to keep an eye on what's changing, we need to monitor them, how might they be affected. We don't need to be ultra hyper-concerned, but we need to monitor the situation."

Mads Peter Heide-Jørgensen agrees, noting that the future health of narwhal populations depends mostly on global warming. "Narwhal abundance worldwide is not very large and it has never been very large. It has probably never exceeded 100,000 animals worldwide," he said. "So because of that, they are a vulnerable species. They have a restricted diet, they mainly eat two or three species, so if something happens to those species, then they may not be able to adapt to those changes. For instance, if halibut is no longer that prevalent on the feeding grounds, then this will harm the narwhals.

"Changes in sea ice conditions can also affect the narwhals," Heide-Jørgensen added. "Since they have adapted to the environment over 10,000 years, if something changes over a period of only a few decades, then of course they may not be capable of adapting. All of these ice entrapment situations are a good example of how lousy they are at adapting to changes in their environment."

The Canadian scientists seemed less willing to speculate about the future, in part because so much is still unknown about the whale and so much is uncertain about future environmental conditions. Pierre Richard said, "I'm not the kind of person who will likely make statements of impending doom, but neither will I want to emphasize having more understanding of the situation than I do."

Richard's colleague, Jack Orr, admitted to having "no crystal ball here, I'm afraid," and then re-emphasized all the challenges that narwhals face. But he concluded our discussion with a tiny note of optimism. "They live in one of the most hostile environments on the planet, having so much darkness in their world, their access to air covered by 98 percent ice most of the year, competition for food with commercial fishing, and all the predators literally wanting a piece of them, including man, of course," he said. "It's a credit to the species to be here at all."

ACKNOWLEDGMENTS

THE RESEARCH AND WRITING OF THIS BOOK WAS QUITE AN adventure—both during the numerous field trips I undertook as well as back home in my office—and none of it could have been completed without the considerable support of many individuals and agencies. I am particularly indebted to Jack Orr at the Canada Department of Fisheries and Oceans, who allowed me to join his research team in Tremblay Sound for two weeks, and to the Office of Research at the University of Rhode Island for its support of my trip to Greenland.

Equally important were the many scientists who shared their insights about narwhals and other wildlife and about the Arctic environment: Pierre Richard, Kristin Laidre, Martin Nweeia, Rune Dietz, Mads Peter Heide-Jørgensen, James Finley, Marie Auger-Méthé, Marianne Marcoux, Ari Daniel Shapiro, Eva Garde, Greg Stone, Fred Eichmiller, Kate Moran, Mark Serreze, Walt Meier, Eric Cravens, Justin Richard, Brigit Braune, Mark Mallory, Brendan Kelly, Gary Stern, Pat Hall, Winston Kuo, Ken Smith, and Robert Park.

In Greenland, my deepest thanks go out to Mads Ole Kristiansen and fellow hunters Gedion Kristiansen and Thomas Qujaukitsoq, as well as to Paaviaaraq Kristiansen, Hans Jensen, and Finn Hansen. My companions in Tremblay Sound, who provided vital guidance and assistance, were Sandie Black, Clint Wright, Joe Guay, Cortney Watt, Jim Orr, Frank McCann, James Simonee, Ronie Komangapik, Josh Idlout, and David Angnetsiak.

Thanks also to Judy Chupasko at the Harvard Museum of Comparative Zoology, Laura Pereira at the New Bedford Whaling Museum Research Library, Alison Rich at the John Carter Brown Library, Annelise Sorg at the Canadian Marine Environment Protection Society, Peter August at the University of Rhode Island's Environmental Data Center, Jacinta Simoncini at Mystic Aquarium, and to Josh Araujo, Kim Robertson, Namen Inuarak, Walt and Ellen Newsom, Kim and Ed Walker, Teresa Gervelis, and my co-workers at the University of Rhode Island's Department of Communications and Marketing.

I also wish to thank my literary agent Charlotte Raymond, whose early enthusiasm for my writing ultimately led to this project, and to the staff at the University of Washington Press, especially Marianne Keddington-Lang, Tim Zimmerman, Beth DeWeese, Marilyn Trueblood, Tom Eykemans, and Rachael Levay.

For their encouragement during the research and writing of this book, I wish to acknowledge my parents Jordan McLeish and Herb McLeish. But most of all, my greatest thanks are reserved for my wife, Renay, who finally was able to join me on some of my research adventures and who contributed to the completion of this book in so many valuable ways. Her undying support for my writing is immeasurable, despite the fact that it means that we are separated far more often than either of us would prefer. Thank you so very much.

BIBLIOGRAPHY

Agreement between the Inuit of the Nunavut Settlement Area and Her Majesty the Queen in Right of Canada, § Article 5. 1993.

Aldrovandi, Ulisse, et al. *Ulyssis Aldrovandi . . . de Piscibus libri V.* Bononiae, 1608.

Alley, Sam. "Knud Johan Victor Rasmussen." Minnesota State University E-Museum. Web.

Anderson, Johann. *Histoire naturelle de l'island du Greenland.* Paris: Michael Lambert, 1750.

"Arctic Whales Assist Scientists Taking Ocean Temperatures." *LiveScience,* 28 October 2010. Web.

Auger-Méthé, Marie, Marianne Marcoux, and Hal Whitehead. "Nicks and Notches of the Dorsal Ridge: Promising Mark Types for the Photo-Identification of Narwhals." *Marine Mammal Science,* December 2009.

Banks, Joseph. *Letters on Iceland: Containing Observations on the Civil, Literary, Ecclesiastical, and Natural History.* London: W. Richardson, 1780.

Bird, John. "Narwhal Tragedy Yields Harvest for Community." *Nunatsiaq (Iqaluit) News,* 28 November 2008.

Black, Richard. "Whales Take Northwest Passage as Arctic Sea-Ice Melts." BBC News, 21 September 2011. Web.

Boaz, Franz. *The Central Eskimo.* Lincoln: University of Nebraska Press, 1964.

Boswell, Randy. "Arctic Ice Melt Threatens the Iconic Narwhal." *Vancouver (B.C.) Sun,* 8 September 2010.

Bowman, Lee. "Unicorn Whales Hooked on a Feeling." *(Toronto) Globe and Mail,* 13 December 2005.

Braune, Birgit, M. L. Mallory, and H. G. Gilchrist. "Elevated Mercury Levels in a Declining Population of Ivory Gulls in the Canadian Arctic." *Marine Pollution Bulletin* 52 (2006): 969–87.

Broad, William J. "It's Sensitive. Really." *New York Times*, 13 December 2005.

Burek, Kathy A., Frances M.D. Gulland, and Todd M. O'Hara. "Effects of Climate Change on Arctic Marine Mammal Eealth." *Ecological Applications* 18, no. 2 (2008): S126–34.

Candela, Phil. "Polar Bear Evolution." University of Maryland, Department of Geology. Web.

Chivers, C. J. "Russia Plants Flag on Sea Floor at North Pole." *New York Times*, 2 August 2007.

COSEWIC. *Assessment and Update Status Report on the Atlantic Walrus Odobenus rosmarus rosmarus in Canada.* Committee on the Status of Endangered Wildlife in Canada, Ottawa, 2006.

Cuvier, Frederick G. *De l'histoire naturelle des Cetaceces.* Paris: Librairie Encyclopedique de Roret, 1836.

Darwin, Charles. *The Descent of Man, and Selection in Relation to Sex.* London: John Murray, 1871.

Davis, T. Neil. "Aurora in Eskimo Legend." *Alaska Science Forum.* Geophysical Institute, University of Alaska Fairbanks. Web.

Desire. "My Passion for Unicorns." *Dragon's Lair,* 25 December 2007. Web.

Dewar, Mirian, ed. *The Nunavut Handbook.* Iqaluit, Nunavut: Ayaya Marketing and Communications, 2004.

Dewhurst, Henry William. *The Natural History of the Order Cetacea, and the Oceanic Inhabitants of the Arctic Regions.* London: H. W. Dewhurst, 1834.

Dickinson, Terence. *NightWatch: A Practical Guide to Viewing the Universe.* Willowdale, ON: Firefly, 1998.

"Did Whale Evolution Go Backwards?" Smithsonian Museum of Natural History. Web.

Dietz, Rune, Ari D. Shapiro, Mehdi Bakhtiari, Jack Orr, Peter L. Tyack, Pierre Richard, Ida Eskesen, and Greg Marshall. "Upside-down Swimming Behaviour of Free-Ranging Narwhals." *BMC Ecology* 7, no. 1 (2007): 14.

Dietz, Rune, F. Riget, K. A. Hobson, M. P. Heide-Jørgensen, P. Moller, M. Cleemann, J. De Boer, and M. Glasius. "Regional and Inter-Annual Patterns of Heavy Metals, Organochlorides and Stable Isotopes in Narwhals from West Greenland." *Science of the Total Environment* 331 (2004): 83–105.

Dietz, Rune, Mads Peter Heide-Jørgensen, Pierre Richard, Jack Orr, Kristin Laidre, and Hans Christian Schmidt. "Movements of Narwhals (*Mon-*

odon monoceros) from Admiralty Inlet Monitored by Satellite Telemetry." *Polar Biology* 31, no. 11 (2008): 1295–1306.

Dietz, Rune, M. P. Heide-Jørgensen, P. R. Richard, and M. Acquarone. "Summer and Fall Movements of Narwhals from Northeastern Baffin Island towards Northern Davis Strait." *Arctic* 54, no. 3 (2001): 244–61.

"Digital Tags Provide Evidence that Narwhals May Produce Signature Vocalizations for Communication." Woods Hole Oceanographic Institute, 28 September 2006. Web.

"DOSITS: Narwhal." *Discovery of Science in the Sea*. University of Rhode Island. Web.

Dunn, Jon, and Jonathan K. Alderfer. *National Geographic Field Guide to the Birds of North America*. Washington, DC: National Geographic, 2006.

Ehrlich, Gretel. *This Cold Heaven: Seven Seasons in Greenland*. New York: Vintage Books, 2003.

Ellis, Henry. *A Voyage to Hudson's-Bay: By the Dobbs Galley and California, in the Years 1746 and 1747, for Discovering a North West Passage*. London: H. Whitridge, 1748.

Ellis, Richard. *Men and Whales*. New York: Lyons Press, 1999.

"Evolution Library: Whale Evolution." Public Broadcasting Service, 2001. Web.

Freeman, Milton M. R., et al. *Inuit, Whaling and Sustainability*. Walnut Creek, CA: AltaMira Press, 1998.

Funk, McKenzie. "Arctic Land Grab." *National Geographic*, May 2009, 104–21.

Gaden, A and Gary Stern. "Temporal Trends in Beluga, Narwhal and Walrus Mercury Levels: Links to Climate Change." In *A Little Less Arctic: Top Predators in the World's Largest Northern Inland Sea, Hudson Bay*, edited by Steven H. Ferguson, Lisa L. Loseto, and Mark L. Mallory, 197–215. Dordrecht, Netherlands: Springer Science+Business Media, 2010.

Gesner, Konrad. *Historiae Animalium*. Zurich: Christof Froshover, 1558.

Gilchrist, H. Grant, and Mark L. Mallory. "Declines in Abundance and Distribution of the Ivory Gull (*Pagolphila eburnea*) in Arctic Canada." *Biological Conservation* 121 (2005): 303–9.

Greenfield-Boyce, Nell. *Chasing after the Elusive Narwhal*. National Public Radio, Washington, DC, 18 August 2009.

Hall, Elizabeth, and Max Hall. *About the Exhibits*. Museum of Comparative Zoology, Harvard University, Cambridge, MA, 1975.

Hamilton, Robert. "On the Ordinary Cetacea or Whales." In *The Naturalist's Library: Mammalia*, vol. 6, 182–90. Edinburgh: W. H. Lizars, 1837.

Harington, C. R. "The Evolution of Arctic Marine Mammals." *Ecological Applications* 18, no. 2 (2008): S23–40.

Heide-Jørgensen, Mads Peter, and Kristin Laidre. *Greenland's Winter Whales: The Beluga, the Narwhal, and the Bowhead Whale.* Ilinniusiorfik, Greenland: Undervisningsmiddelforlag, 2006.

"The History of Greenland." Official Greenland Travel Guide. Web.

Hume, Mark. "A Bittersweet Harvest from a Misfortune of Nature." *(Toronto) Globe and Mail,* 26 November 2008.

Ice Islands. DVD. Directed by Greg Stone. National Geographic Society, 2003.

"Inuit Legend." *Narwhal Tusk Discoveries.* Web.

"Inuit Seek Review of Narwhal Tusk Trade Ban." CBC News, 11 January 2011.

Jackson, Sophie. *The Horse in Myth and Legend.* Gloucestershire, U.K.: Tempest Publishing Ltd., 2006.

Johnson, Mark P. "Playing Tag with Whales." *Oceanus,* Woods Hole Oceanographic Institute, March 2005.

Joling, Dan. "Warming Means Ringed Seals Face an Uncertain Future." *Anchorage Daily News,* 13 December 2010. Web.

Kavenna, Joanna. *The Ice Museum: In Search of the Lost Land of Thule.* New York: Viking, 2006.

Kelly, Brendan. "Climate Change and Ice Breeding Pinnipeds." In *Fingerprints of Climate Change: Adapted Behavior and Shifting Species Ranges,* edited by G.-R. Walther, C. A. Burga, and P. J. Edwards, 43–56. New York: Kluwer Academic, 2001.

Kelly, Brendan, Andrew Whiteley, and David Tallmon. "The Arctic Melting Pot." *Nature* 468 (2010): 891.

"Killing Pond Inlet Narwhals 'Humane Harvest': DFO." CBC News, 24 November 2008.

Kingsley, Michael, and Pierre Richard. "Underwater World—Narwhal." Fisheries and Oceans Canada, 2007. Web.

Kunz, George F. *Ivory and the Elephant: In Art, in Archaeology, and in Science.* New York: Doubleday, Page and Co., 1916.

Laidre, Kristin L., and Mads Peter Heide-Jørgensen. "Arctic Sea Ice Trends and Narwhal Vulnerability." *Biological Conservation* 121 (2005): 509–17.

Laidre, Kristin L., Ian Stirling, Lloyd Lowry, Oystein Wiig, Mads Peter Heide-Jørgensen, and Steven Ferguson. "Quantifying the Sensitivity of Arctic Marine Mammals to Climate-Induced Habitat Change." *Ecological Applications* 18, no. 2 (2008): S97–125.

Laidre, Kristin, M. P. Heide-Jørgensen, and J. R. Orr. "Reactions of Narwhals, *Monodon monoceros*, to Killer Whale, *Orcinius orca*, Attacks in the Eastern Canadian Arctic." *Canadian Field Naturalist* 120 (2006): 457–65.

Laing, John. *An Account of a Voyage to Spitzbergen; Containing a Full Description of that Country, of the Zoology of the North and of the Shetland Isles; with an Account of the Whale Fishery.* London: L. Mawman, 1815.

La Martiniére, Pierre M. *Nouveau voyage du nort.* Amsterdam: Estienne Roger, 1700.

La Peyrère, Isaac. *Relation du Groenland.* Paris: Augustin Courbe, 1663.

Lavers, Chris. *The Natural History of Unicorns.* New York: William Morrow, 2009.

Lopez, Barry. *Arctic Dreams: Imagination and Desire in a Northern Landscape.* New York: Charles Scribner's Sons, 1986.

Lord, Nancy. *Beluga Days: Tracking the Endangered White Whale.* Seattle: Mountaineers Books, 2007.

Magnus, Olaus. *Historia de gentibus septentrionalibus.* Rome: De Viottis, 1555.

Manby, George W. *Journal of a Voyage to Greenland, in the Year 1821.* 2d ed. London: G. and W. B. Whittaker, 1823.

Marcoux, Marianne. "Social Behaviour, Vocalization and Conservation of Narwhals." *InfoNorth*, 456–60.

Marcoux, Marianne, and Marie Auger-Méthé. "Encounter Frequencies and Grouping Patterns of Narwhals in Koluktoo Bay, Baffin Island." *Polar Biology* 32 (2009): 1705–16.

Melville, Herman. *Moby-Dick, or, The Whale.* London: Harper Brothers, 1851.

Moore, Sue E., and Henry P. Huntington. "Arctic Marine Mammals and Climate Change: Impacts and Resilience." *Ecological Applications* 18, no. 2 (2008): S157–65.

Moran, Kathryn, and Jan Bachman. "A Cenozoic History of the Arctic Ocean." *Oceanography* 19, no. 4 (2006).

Mulvaney, Kieran. "Loss of Sea Ice Poses Mercury Risk." *Discovery News*, 26 January 2011. Web.

Mulvaney, Kieran. "Polar Bears Found Rock Climbing in the Arctic." *Discovery News*, 26 April 2010. Web.

———. "'Unicorn' Whales Track Arctic Temperatures." *Discovery News*, 28 October 2010. Web.

Neme, Laurel A. *Animal Investigators: How the World's First Wildlife Forensics Lab is Solving Crimes and Saving Endangered Species*. New York: Charles Scribner's Sons, 2009

Neruda, Pablo. *Memoirs*. London: Penguin, 1998.

Newman, Murray. *Life in a Fishbowl: Confessions of an Aquarium Director*. Vancouver, B.C.: Douglas & McIntyre, 1994.

Nicklen, Paul. "Arctic Ivory." *National Geographic*, August 2007, 110–29.

———. *Polar Obsession*. Washington, DC: National Geographic, 2009.

"Nunavut Territory Map." Nunavut Tourism, 2007.

Nweeia, Martin T., et al. "Considerations of Anatomy, Morphology, Evolution and Function for Narwhal Dentition." In *Smithsonian at the Poles: Contributions to International Polar Year Science*. Washington, DC: Smithsonian Institution Scholarly Press, 2008.

O'Corry-Crowe, Gregory. "Climate Change and the Molecular Ecology of Arctic Marine Mammals." *Ecological Applications* 18, no. 2: S56–76.

O'Reilly, Bernard. *Greenland, the Adjacent Seas, and the Northwest Passage to the Pacific Ocean*. London: Baldwin, Cradock, and Joy, 1818.

Osborn, Liz. "Ivory Gulls Mysteriously Disappear." *Current Results*. Web.

Paré, Ambrose. *Discours d'Ambroise Paré*. Paris: Gabriel Buon, 1582.

"Petition to List Three Seal Species under the Endangered Species Act." Center for Biological Diversity, 28 May 2008

Pielou, E. C. *A Naturalist's Guide to the Arctic*. Chicago: University of Chicago Press, 1994.

Pollack, H. N. *A World without Ice*. New York: Avery, 2009.

Ragen, Timothy J., Henry P. Huntington, and Grete K. Hovelsrud. "Conservation of Arctic Marine Mammals Faced with Climate Change." *Ecological Applications* 18, no. 2 (2008): S166–74.

Rasmussen, Knud. *Across Arctic America: Narrative of the Fifth Thule Expedition*. New York: G. P. Putnam's Sons, 1927.

Reeves, Randall, Brent Stewart, Phillip Clapham, and James Powell. *National Audubon Society Guide to Marine Mammals of the World*. New York: Alfred A. Knopf, 2002.

Revkin, Andy. "Arctic Once Felt like Florida, Studies Say." *New York Times*, 31 May 2006.

———. "Hints of Oil Bonanzas beneath Arctic Ocean." *New York Times*, 1 June 2006.

———. "Walruses Suffer Substantial Losses as Sea Ice Erodes." *New York Times*, 3 October 2009.

Richard, P. R., J. L. Laake, R. C. Hobbs, M. P. Heide-Jørgensen, N. C. Asselin, and H. Cleator. "Baffin Bay Narwhal Population Distribution and Numbers: Aerial Surveys in the Canadian High Arctic, 2002–2004." *Arctic* 63, no. 1 (2010).

Ross, John. *A Voyage of Discovery, Made under the Orders of the Admiralty in His Majesty's Ships* Isabella *and* Alexander, *for the Purposes of Exploring Baffin's Bay*. London: John Murray, 1819.

Rubin, Jeff. "Life on Ice." *Audubon*, January–February 2009, 51–55.

Sale, Richard. *A Complete Guide to Arctic Wildlife*. Buffalo, NY: Firefly Books, 2007.

"Scientist Fears Disease Outbreaks in Northern Whales—North." CBC News, 18 August 2006.

Scoresby, William. *An Account of the Arctic Regions, with a History and Description of the Northern Whale Fishery*. Edinburgh: A. Constable, 1820.

Service, Robert W. *Collected Poems of Robert Service*. New York: Putnam, 1989.

Shepard, O. *The Lore of the Unicorn*. London, 1930.

Silis, Ivars. "Narwhal Hunters of Greenland." *National Geographic*, April 1984, 520–39.

Simmonds, Mark P., and Wendy J. Eliott. "Climate Change and Cetaceans: Concerns and Recent Developments." *Journal of the Marine Biological Association of the United Kingdom* 89 (2009): 203–10.

Smith, G. *The Entertaining Correspondent, or Curious Relations, Digested into Familiar Letters and Conversations*, vol. 2. London: James Hodges, 1739.

Smith, Kenneth, and Bruce H. Robison. "Free Drifting Ice Bergs." *Science* 27, no. 317 (2007): 478–82.

Sperling, Johannis. *Zoologia physica posth*. Witteberg: Johannis Bergeri, 1661.

Stone, Greg. "Exploring Antarctica's Islands of Ice." *National Geographic*, December 2001.

———. *Ice Island: The Expedition to Antarctica's Largest Iceberg*. Boston: Bunker Hill, 2003.

Streever, Bill. *Cold: Adventures in the World's Frozen Places*. New York: Little, Brown, 2009.

Swartz, Spencer. "Oil Exploration Expands in Greenland." *Wall Street Journal*, 25 April 2010.

Vastag, Brian. "Shifting Spring: Arctic Plankton Blooming up to 50 Days Earlier Now." *The Washington (DC) Post*, 7 March 2011.

Verne, Jules. *Twenty Thousand Leagues under the Sea.* New York: Dodd, Mead, 1952.

Walsh, John E. "Climate of the Arctic Marine Environment." *Ecological Applications* 18, no. 2 (2008): S3–22.

Wörm, Olaus. *Museum Wormianum.* Lyon: Johannem Elsevirium, 1655.

Zeitvogel, Karin. "Arctic Sea Ice Melts to Near-Record Level." *Discovery News*, 16 September 2010. Web.

INDEX

A

Academy of Natural Sciences of Philadelphia, 137

Admiralty Inlet, 5, 60, 151, 171

Aelianus, Claudius, 63

aerial survey, 25–26, 29, 32, 107, 138

Alaska Veterinary Pathology Services, 95

albedo, 87, 92

Aldrovandi, Ulisse, 12

algal bloom, 23, 93, 95, 180

alicorn, 65

Ambulocetus, 80

American Dental Association Foundation, 43

Angnetsiak, David, 120–21

Antarctica, 21–23, 86

Antarctic krill, 21–23

Archaeocetes, 80

Arctic Bay, 114, 156

Arctic char, 37

Arctic cod, 24, 93, 99

Arctic poppy, 161

Arctic willow, 17

ARGOS, 33

Aristotle, 63

assallut, 103

Auger-Méthé, Marie, 165–70

aurora borealis, 183–85

awatuk, 102–3, 112

Azolla, 85–86

B

Backman, Jan, 84

Baffin Bay, 4, 16, 69, 141; aerial survey of, 107; algal blooms in, 95; and bowhead whales, 177–78; and fishing, 145; and narwhal wintering, 29, 81, 177; sea ice, 93; temperatures, 124–25; and walrus, 46

Baffin Island, 5–8, 16, 24, 35–36; and bowhead migration, 179; and ivory gulls, 129–30

Barren Grounds, 6

Basilosaurus, 80

Beaufort Sea, 157

beluga whale, x, 18, 29, 60, 83, 137, 155; in captivity, 150, 156–60; feeding, 93, 109; hunting of, 113; hybrid of, 83, 97; and pollutants, 143; tagging of, 146

Bering Sea, 157, 180

Black, Sandie, 152, 155, 175–76

Blossville Coast, 138

blue whale, 113

Bone Cell Biology Lab, 41

Boston Children's Hospital, 41

E

East Greenland, 29, 81, 125; hunting quotas in, 125; ice entrapments in, 109; narwhal research in, 138–41
echolocation, 24, 160, 162, 171, 174, 177, 180
Eclipse Bay, 9, 21
Eclipse Sound, 5, 8, 58, 147, 177
Ehrlich, Gretel, 103, 122, 132, 148
Eichmiller, Fred, 43–45, 50–53
elephant, 41
Ellesmere Island, 100, 121, 130, 156
Ellis, Henry, 13
Ellis, Richard, 66
enamel, tooth, 44–45, 50–51
England, 81, 85
Environment Canada, 130
Erik the Red, 123
Eriksson, Leif, 35

F

Finley, James K., 28–29, 55–56, 117–19
fin whale, 113, 135
firn, 90
fishing: by Inuit, 21, 99; commercial, 83, 139, 145, 187–88
Flaherty, Robert, 36
flensing, 127
Fønfjord, 138
food chain, 23, 93, 130, 141, 143
Foxe Basin, 46, 60, 95, 179
Franklin, John, 36, 136
Franz Josef Land, 138
Freeman, Milton, 115–16
Frobisher, Martin, 35
Frobisher Bay, 7, 35–36
fulmars, northern, 9–10, 18–19, 21, 30, 99, 111, 127–28

G

Garde, Eva, 181–83
geology, Arctic Ocean, 85
Gesner, Konrad, 11
glacier, 22, 81, 87–90, 96, 97, 133
glaucous gull, 9, 38, 99, 129
global oceanic conveyor belt, 139–40
global warming. *See* climate change
Globe and Mail (Toronto), 121
Graduate School of Oceanography, University of Rhode Island, 84
greenhouse gases, 85–86, 91–92
Greenland, history, 122–24
Greenland halibut, 24, 107, 145, 151, 177, 188
Greenland Institute for Natural Resources, 181
Greenland Sea, 139–40
Greenland Summit, 88
Grise Fjord, 156
Guay, Joe, 26, 154
Gulf of St. Lawrence, 45

H

Hamilton, Robert, 15
Harper, Kenn, 35
harpoon, x, 74, 101–3, 112–14, 126–27; and bowhead whale, 178–81; use in ice entrapment, 120; use in research, 105–6
harp seal, 10
Harvard Museum of Natural History, 135
Harvard University, 40, 135
Heide-Jørgensen, Mads Peter, 105, 126, 137–40, 179, 187–88
Helluland, 35
Hudson Bay, 7, 24, 29, 33, 60, 113, 157; walrus population in, 45–46

long-tailed jaeger, 31
Lopez, Barry, 5, 20, 115, 178, 184

M

magnetic north, 100
Magnus, Olaus, 11
Mallory Mark, 130
manatee, 172
Marcoux, Marianne, 164–70
Marine Mammal Protection Act, 40
Massachusetts Institute of Technology, 53
mattak, 115, 126. *See also* muktuk
McCann, Frank, 163
McGill University, 164
Meier, Walt, 133–34
Melville, Herman, 52
Melville Bay, 107, 124–26
Mercator, Gerhard, 65
mercury contamination, 116, 130, 140–42
Meta Incognita peninsula, 7
Milne Inlet, 10, 19, 58
minke whale, 108, 114
Mittimatalik, 37. *See also* Pond Inlet
Monodon monoceros, 4, 11
Monterey Bay Aquarium Research Institute, 21
Moran, Kate, 84–86, 90–91
moss campion, 161
mountain sorrel, 16
muktuk, 113, 115–16, 118, 122, 126–27, 131–32, 178
musk ox, 19, 102, 122
Mystic Aquarium, 158

N

narwhal(s): aging of, 181–82; behavior of, 20, 54, 107, 153–54, 169–70, 173; in captivity, 155–57; collecting water temperatures via, 124–25, 139–40; communication among, 14, 170–73; diseases in, 95, 120; diving abilities of, 32–33, 107, 177; evolution of, 42, 80–81, 118, 182; fetus of, 42; food of, 24; future outlook for, 187–88; genetics of 29, 42, 81, 137, 182; health of, 95, 120, 143–44, 155; identification of individual, 166, 169; management of, 105, 114, 150; migration of, 4–5, 81, 107, 139; physiology, 81–82; populations of, 4, 24–25, 28–29, 81, 107, 116–17, 125–26, 138, 151; predators of, 60–61, 176; range of, 4,29, 81, 137–39; swimming, 96, 173–74; tagging of, 33, 106, 139, 149, 151, 174–76; vocalizations of, 30, 162, 166, 170–73. *See also* narwhals, hunting of; tusks, narwhal
narwhals, hunting of, 74, 100–104, 110–20, 125–32, 185; camp for, 6, 104, 110–11; in Canada, 114, 118–20; and Inuit culture, 115–16; methods of, 101, 110, 114; opposition to, 116–18; quotas for, 114, 125–26, 138. *See also* whaling
Narwhalus, 15
National Environmental Research Institute, 141
National Geographic magazine, 118
National Geographic Society, 173
National Ice Core Laboratory, 87
National Oceanic and Atmospheric Administration, 150
National Public Radio, 171
National Snow and Ice Data Center, 91, 133
Navy Board Inlet, 177
Neruda, Pablo, 4

raven, 9

red-breasted merganser, 18

reem, 63

Repulse Bay, 61, 114, 179

Resolute Bay, 147

rhinoceros, 63

Rhode Island, 3, 6, 61

Rhodocetus, 80

Richard, Justin, 159–60

Richard, Pierre, 24–29, 32–33, 106, 188

Richards, Mike, 121

right whale, 135–36, 172

ringed seal, 9–10, 19, 37, 83

Rosenberg Castle, 65

Ross, John, 36

Ross Ice Shelf, 22

Russia, 4, 84–85, 100, 137–38; and bowhead whales, 179; and ivory gulls, 130–31; and walruses, 46–47

S

Sabine's gull, 58

Safina, Carl, 30

Saqqaq people, 122

satellite tags, 60, 105, 139, 149, 176–77

savssats. See ice entrapments

sawfish, 11–12

saxifrage, 17

School of Dental Medicine, Harvard University, 40

Scoresby, William, 14, 54, 56

Scoresby Sound, 138–39

Scripps Institution of Oceanography, 95

sea cucumber, 23

sea ice, 3, 8, 69, 132–34; and bowhead whales, 180; changing patterns of, 95, 121, 182; and climate change, 92–93, 182; and fishing, 145; formation, 133; in history, 85–87;

and ivory gulls, 130–31; and killer whales, 61; and narwhals, 93–94, 107, 182, 188; and polar bears, 82–83; and pollutants, 142–43; types of, 133–34; and walruses, 47–48

sea lice, 175

seals: harp, 10; ringed, 9–10, 19, 37, 83

Sermilik fjord, 138

Serreze, Mark, 91–93

Service, Robert, 184

Seymour Island, 130

Shapiro, Ari Daniel, 171–73

Shepard, Odell, 63–64

shipping, 3, 83, 94, 144, 180

Siorapaluk, 121

Siunnertalik, 102, 104–5, 127–28. *See also* narwhal, hunting of

Smith, G., 13

Smith, Ken, 21–23

Smith Sound, 121

Sorg, Annelise, 117

Southern Ocean, 22–23

sperm whale, 113, 135–36, 172–73

Spitsbergen, 179

St. Lawrence River, 32, 143, 157–58

Stern, Gary, 143

Stockholm University, 84

Stone, Greg, 22–23

Streever, Bill, 88

Survivorman, 16, 59

swimming, 96. *See also* upside-down swimming

T

Tadoussac, 158

Tasiilaq, 125, 138

thule, 123–24

Thule Air Base, 123

Thule people, 122–24

Tibetan antelope, 63

Ticuna Indians, 41

Tremblay Sound, 147, 151, 153, 174, 177–78, 186

tusking behavior, 153–54, 162, 170. *See also* narwhals, behavior of

tusks, narwhal, 3–4, 31–34, 37–39, 40–45, 49–57, 72, 175; in archaeology, 137; hunting, 114–18, 126–27; evolution of, 42, 49, 54, 118; genetic analysis of, 137; historic depictions of, 12–15, 67; medicinal properties of, 63, 66; pollutants in, 141–42; purpose of, 33–34, 52–57; sale of, 40, 114, 117; spiral of, 41, 43–44, 174; structure of, 40–45; tubules in, 50–51, 53; and unicorns, 63–66

U

Uganda, 66

unicorn, 3, 11, 15, 61–65

Unicorn Tapestries, 64

University of Alaska, 47

University of Alberta, 165

University of California at Berkley, 89

University of California at Santa Cruz, 96

University of Colorado, 91

University of Copenhagen, 181

University of Manitoba, 26, 143

University of Nottingham, 63

University of Rhode Island, 84

University of Washington, 54, 104

upside-down swimming, 106, 167, 173

U.S. Geological Survey, 47, 87, 144

Uummannaq, 107

V

Vancouver Aquarium, 156–57

Verne, Jules, 15

Vikings, 35, 64, 66

"Vletif," 12

W

Wall Street Journal, 144

walrus, ix, 45–48, 150, 183; and climate change, 47–48, 83; and contaminants, 143; hunting of, 46, 101, 118; populations of, 45–46; stampedes of, 46–47; tusks, 46, 181

Watt, Cortney, 154

Weddell Sea, 21

West Greenland, 4–5, 29, 81, 137, 139–41; bowhead whales in, 177, 179; killer whales in, 60; narwhal hunting in, 125–26; narwhal populations in, 107

Whale and Dolphin Conservation Society, 117

whale culture. *See* cultural hitchhiking

whale meat, and health, 116

whales, evolution of, 80–81, 108, 182; fossils of, 41, 81. *See also specific whale species*

whaling, 11–12, 36, 113, 115–17, 179

Whitehead, Hal, 166

Wildlife Computers, 33

Williams, Terrie, 96

Woods Hole Oceanographic Institute, 53, 171

World Wildlife Fund, 47

Wörm, Olaus, 12, 65

Wright, Clint, 155–57

Y

Yap Island, 41

Lightning Source UK Ltd.
Milton Keynes UK
UKHW010029200320
360651UK00001B/603

9 780295 997353